UX Writing

This flexible textbook provides an integrated approach to user experience (UX) writing and equips students and practitioners with the essential principles and methods to succeed in writing for UX.

The fundamental goal of UX writing is to produce usable and attractive content that boosts user engagement and business growth. This book teaches writers how to create content that helps users perform desired tasks while serving business needs. It is informed by user-centered design, content strategy, artificial intelligence (AI), and digital marketing communication methodologies, along with UX-related practices. By combining writing-as-design and design-as-writing, the book offers a new perspective for technical communication education where UX design and writing are merged to achieve effective and desirable outcomes.

Outlining the key principles and theories for writing user-centered content design, this core textbook is fundamental reading for students and early career practitioners in UX, technical communication, digital marketing, and other areas of professional writing.

Jason C. K. Tham is an Associate Professor of Technical Communication and Rhetoric at Texas Tech University, USA.

Tharon Howard is a Professor of English at Clemson University, USA.

Gustav Verhulsdonck is an Associate Professor of Business Information Systems at Central Michigan University, USA.

UX Writing

Designing User-Centered Content

Jason C. K. Tham, Tharon Howard, and
Gustav Verhulsdonck

Routledge
Taylor & Francis Group

NEW YORK AND LONDON

Designed cover image: solvod / © Getty Images

First published 2024
by Routledge
605 Third Avenue, New York, NY 10158

and by Routledge
4 Park Square, Milton Park, Abingdon, Oxon, OX14 4RN

Routledge is an imprint of the Taylor & Francis Group, an informa business

© 2024 Jason C. K. Tham, Tharon Howard, and Gustav Verhulsdonck

The right of Jason C. K. Tham, Tharon Howard, and Gustav Verhulsdonck to be identified as authors of this work has been asserted in accordance with sections 77 and 78 of the Copyright, Designs and Patents Act 1988.

ISBN: 9781032228389 (hbk)
ISBN: 9781032227405 (pbk)
ISBN: 9781003274414 (ebk)

DOI: 10.4324/9781003274414

Typeset in Sabon
by codeMantra

Contents

Foreword

Louisiana Tech University and the University of Limerick

When I first learned of this new textbook on UX writing, I was pleased for a number of reasons. The first was that this collaborative work resulted from the Louisiana Tech Usability Studies Symposium (LATUSS)—the event where the authors first met and first began to discuss these ideas. Second, the book represents the realization of LATUSS's focus on "rethinking usability" by helping readers reconsider the role usability plays in design thinking, content strategy, and technical writing as these areas relate to the burgeoning field of UX writing. Finally, and perhaps most importantly, the book does a masterful job of advancing LATUSS's focus on cognition in communication—particularly in how the authors provide creative solutions to what I call the "3Cs of cognition."

Usability is a matter of cognition or how the human mind processes information. Essentially, the cognitive dynamics that affect how we perceive our surroundings influence the way we use items in an environment. Such cognitive connections are not innate. Rather, they are shaped by our experiences over time, and these dynamics play a foundational role in the complexity of usability. Moreover, they often involve three interrelated factors, the three Cs: Conditioning, Context, and Content. The better we can identify, understand, and address these 3C aspects, the more effectively we can design items individuals find "usable" based on their experiences. Understanding such factors is central to integrating usability into more traditional technical writing processes.

Conditioning encompasses how we learn to use items based on our experiences and can generally involve three approaches: (1) **Active Instruction**: when someone actively teaches us how to use an item; (2) **Passive Observation**: learning from others and mimicking their behavior; and (3) **Independent Interaction**: when we learn by "playing" with a new item/ technology.

In today's technological environments, however, technical writers need to increasingly rethink the nature of conditioning as related to usability. This is because the increase in "out-of-the-box" UX now leads to a growing demand for independent interaction. Simply put, users want *to immediately do things* with a product, service, or design, and they want to do so without having to read a manual. This situation brings with it expectations that the related experience is pleasant, memorable, and personalized—all of which go beyond more conventional approaches to usability. This situation means usability is now expected and framed in terms of overall UX. As a result, an understanding of UX factors can help us better identify and address the usability and associated design expectations of different user groups. Such factors can also help us better understand the dynamics of groups interacting in interconnected and wirelessly networked environments.

These conditioning situations are all closely connected to the second C of *context* or the *location* where individuals perform activities. This is because we don't use tools and technologies randomly, nor do we always use them ubiquitously. Rather, we often associate their use (i.e., how to use them) with specific settings. Many of us, for example, are now comfortable using our mobile devices in public spaces. This situation, in turn, has led to different social expectations for meetings, as people can now change their meeting location on the fly by using their mobile devices to send their location to others.

This factor brings us to the final C of *content*. Content encompasses how we design information so individuals can use it to achieve an objective in a setting. Accordingly, for textual/verbal content to be usable, it needs to address what individuals expect to do and use in a specific location. In today's mobile and interconnected contexts, a growing challenge involves addressing the growing need for "microcontent" (small snippets of information *specifically tailored* to an individual user) in a way that meets expectations for creating reusable content. For example, it is common now to ask for the weather and have a digital voice assistant such as *Siri* or *Alexa* provide us with an answer. Such new content expectations also extend to overall product design and affect whether individuals can recognize a tool or technology in order to use it in a given setting.

The difficulty of such situations is that expectations for what constitutes usability have shifted to the broader goal of achieving better UX. The result is a greater demand for positive experiences that are not only usable but also user-friendly, pleasant, and personalized to the user's personal history. When combined with the challenges associated with creating microcontent and meeting the needs of networked users, the complexities of producing usable technical writing can seem overwhelming.

The authors of this book, however, have found a new approach to addressing this situation. It involves rethinking UX writing as a combination of technical writing, design thinking, content strategy, and UX design. It is an approach that both adapts and combines existing areas of study into a method for mapping the conditioning, context, and content factors affecting usability expectations. What makes this approach particularly powerful is that it also addresses usability in terms of the experience surrounding how we use products, services, and overall designs. In addressing such factors, the book offers a new and important perspective to academic researchers and industry practitioners working in UX, content strategy, and design.

To examine these issues, the authors focus on the idea that understanding UX design today means technical writers need to develop new skills—including those associated with design thinking and content strategy. The authors also seek to refocus technical writing practices to include new data analytics technologies—those that gather real-time data on what the user is doing with a technology *while they are using it*.

Rather than addressing the question of "Can we identify the reflexive behavior shaping an individual's usability expectations?" the authors focus on the more fundamental, interconnected questions of "What factors have shaped UX to create such behaviors?" and "How can common industry practices help create the use and design of a broad range of human experiences?" To answer these questions, the authors have written chapters that examine seemingly different areas of design thinking, content strategy, and UX design. These entries, however, are organized and connected in a way that forms a coherent and easily applicable approach to usability in the context of UX. The result is a UX writing approach that is adaptive to different groups, settings, and points in time. It is an approach that identifies the cognitive dynamics shaping user expectations while also addressing industry expectations for incorporating iterative design thinking techniques. In so doing, the approach uses content strategy to develop and manage content. Additionally, it presents UX research strategies to help engage and study how users are interacting with one's content.

By interconnecting such factors, the text represents an important approach to addressing many of the complex topics faced by society today. Moreover, it does so by mixing academic research with industry application in a way that connects to educational practices. As a result, this book can easily be used within academic, industrial, or educational spheres as well as across them. This adaptability makes the text a reference resource and a mechanism for generating new knowledge, novel approaches, and original ideas around the core topic of UX writing. Moreover, the adaptive nature of the approach presented in the text positions it well to stand the test of time.

The better we understand how experiences and exposure shape usability expectations, and vice versa, the more effectively we can create tools and technologies for each other. Likewise, by understanding how cognitive factors of conditioning, context, and content shape such expectations, we can better contribute to society on local, regional, and global levels. Ideally, such a process is a lifelong one as we learn from, think about, and adapt to changing situations that shape our expectations for how to interact with the world around us.

Preface

UX writing—writing for UX—may be new terrain for many readers who picked up this book. You may see UX writing as a growing discipline that promises exciting trajectories and opportunities for seasoned communicators or designers. Others may see UX writing as the merging of technical writing, UX design, and content management—a sign of disciplinary growth. At any rate, UX writing is becoming an expected skill for those who work with digital products and user-facing communication.

As academics who keep our fingers on the pulse of the latest industry trends and innovations, we are excited to share what we've learned from our colleagues and on-the-job practitioners who are leading the curve in establishing field standards and expectations for writing and designing UX. By combining *writing-as-design* and *design-as-writing*, we aim to break new ground for technical communication education where UX design and writing share common missions and values toward social good. This book is an attempt to theorize user-centered content design and apply time-tested, cardinal principles for writing user-centered content. In doing so, we hope to create a new space for technical writing and communication practitioners to expand their expertise and result in positive and desirable outcomes in content design.

Preview of the Book

We have partitioned this book into three major parts. In **Part 1: Perspectives,** we introduce the profession and principles of UX writing by discussing its relationship with technical writing, UX design, usability research, and product development. If you are new to the world of UX, you may appreciate how **Chapter 1** gives you a succinct overview of its influence on current technical communication practices. In conjunction with this introduction, we unveil the emergence of UX writing as a profession and offer a summary of its objectives. In **Chapter 2**, we explain the design thinking model that powers the UX writing process. You will learn the key principles, values, and phases of design thinking as they relate to the content lifecycle. Then, closing out Part 1, **Chapter 3** specifies the desirable traits of those who perform UX writing—preferred skills and expertise—and qualitative attributes for a successful career.

Next, **Part 2: Processes** zeros in on the five core phases of design thinking, starting with **Chapter 4,** where we emphasize the importance of empathy and how it manifests in the practice of user research. After learning about ways to learn about people's attitudes and behaviors with content, you can find methods for translating user needs into actionable insights for content design in **Chapter 5**. In **Chapter 6**, you will learn and exercise ideating content solutions and making them into tangible, testable forms via prototyping. **Chapter 7** teaches you how to test prototyped content and manage its deployment to the public. Then, **Chapter 8** covers various technical and analytical tools to track content performance once it is live.

In the last section of the book, **Part 3: Practices,** we explain the common forms of UX writing and give you guidance on creating your own portfolio. **Chapter 9** goes over six popular UX writing genres and their associated tasks and challenges. You will examine the forms, structures, and delivery of these content types. In **Chapter 10,** you can expect to learn tips and tools for building an attractive UX writing portfolio with confidence. We will show you how to apply the lessons from this book to land a career in UX writing. The final two chapters, **Chapters 11** and **12,** offer an in-depth discussion about generative AI and provide guidance on using it to augment UX writing.

Intended Users

We have written this book with the following audiences in mind:

- **Undergraduate students** in introductory professional writing and technical communication courses as well as special topics seminars in UX writing, usability research, and user-centered design.
- **Instructors** who need pedagogical materials to deliver UX writing assignments that align with technical communication objectives.
- **Graduate students** who require resources for their seminar papers, comprehensive exams, and dissertations.
- **Scholars and researchers** in technical communication who desire an introduction to a new area of research building upon UX, writing, and design theories

This book is both a textbook and a playbook. It can be used to teach a course; it can also be a practical guide to evaluate existing designs and create new solutions. Students may find this book to be an introductory resource toward a UX writing future, whereas practicing professionals may benefit from the exemplary models included in the book.

UX writing combines user-design-centric methods and philosophies that attend to UX, usability, and business objectives. UX writing is integral to product and service design and integrates strategic content deployment. The fundamental goal of UX writing is to produce usable and attractive content that boosts user engagement and business growth. As UX writers draw upon various skills with agility and flexibility to address different work cycles, this book conceptualizes these processes. Our approach teaches writers to create content that helps users perform desired tasks while serving business needs. We strive to help writers develop expertise at the intersection of user research, problem and opportunity definition, content development and management, and continuous iterative design.

Designed to be a flexible core resource, this book offers students and practitioners the essential principles and methods to succeed in writing for UX. Through the perspectives of design thinking, content strategy, user-centered design, and data analytics, this book provides an integrated approach that leads readers into the exciting work of UX writing.

Acknowledgments

Four years felt like a lifetime ago. If you asked us back then in the autumn of 2019—when the three of us met for the first time at a usability studies symposium in Shreveport, Louisiana—whether we would write a book on UX writing, the answer would be a resounding *yes!*

Why, you asked? First, we have Kirk St.Amant to thank for igniting a passion for UX in the field of technical communication and for instilling in the three of us the trust to pull together a project that would wind up being a sustained discussion about modern UX, a series of journal articles, conference presentations, and then... this book! We are especially grateful to Kirk for his leadership and willingness to pen the foreword for this book.

Second, we had a hunch that UX is morphing into a whole new craft that requires specific attention to what technical communication folks have always considered foundational. We thank Ginny Redish for her insights and involvement with this project. Ginny has appeared as a guest lecturer in Tharon and Jason's classes to teach aspirational UX practitioners about designing experiences. More importantly, she has helped provide comments and edited early versions of this book, which made this current iteration the strongest version yet, in our opinion!

Our confidence also came from the field, fueled by those who have taught UX and writing, like Tracy Bridgeford and Rebekka Andersen. We are grateful for their reading of our manuscript and for providing helpful feedback and endorsements.

Of course, producing a book is no solo effort. We are grateful for the support given to us by Routledge via former acquisition editor Brian Eschrich and current editorial assistant Sean Daly. Sean has been especially instrumental in helping us through the manuscript submission and production processes. We must also thank Wendy Howard for volunteering to proofread our book manuscript in the early stages of production. She has a sharp eye for details!

The three of us are also thankful for the support given to us at our respective institutions and through personal connections.

Jason would like to thank his colleagues at Texas Tech University for mentoring him and encouraging his research trajectory. He is also grateful toward his students in UX research and design courses, where he got to learn from them the most current design challenges and immediate UX problems. He also thanks his colleagues in the field who have shared cutting-edge ideas at various venues like the ACM Special Interest Group on Design of Communication, the Council for Programs in Technical and Scientific Communication, and the Association for Teachers of Technical Writing. He is grateful to Tharon and Gustav for sharing their leadership in this project and teaching him all the new things in content strategy and emerging technologies.

Tharon would like to thank his students in the Usability Testing Methodologies graduate seminars, in UX Research seminars, and in the Content Strategy courses for all the examples they created, for giving their permission to use their work in this book, and for helping him see so many different and useful perspectives on UX writing. He would also like to thank his colleagues in the UTEST-L online community who have shared their professional insights into industry practices and the evolving nature of usability and UX research and design over the past 30 years. Thanks also to alumni from the Clemson MAPC and RCID programs now working in industry, who also shared their practitioner insights and their struggles creating UX writing positions and departments in their corporations. Tharon would also like to thank Jason and Gustav for their patience and for putting up with our team's official "old fart" telling war stories about the ways we did things back in the 80s and 90s. Thanks for forcing me to think in 21st-century terms about the future of our profession.

Gustav would like to thank his colleagues at Central Michigan University, colleagues who he has worked with at the Digital Life Institute, met at various conferences, and collaborated in writing different articles. He also wants to acknowledge Jason and Tharon for being awesome collaborators who continuously push him to innovate in his thinking, research, and writing.

Part 1

Perspectives

We open with an introduction to the profession and principles of UX writing by discussing its relationship with technical writing, user experience design, usability research, and product development. If you are new to the world of UX, you may appreciate how **Chapter 1** gives you a succinct overview of its influence on current technical communication practices. In conjunction with this introduction, we unveil the emergence of UX writing as a profession and offer a summary of its objectives. In **Chapter 2**, we explain the design thinking model that powers the UX writing process. You will learn the key principles, values, and phases of design thinking as they relate to the content lifecycle. Then, closing out Part 1, **Chapter 3** specifies the desirable traits of those who perform UX writing—preferred skills and expertise and qualitative attributes for a successful career.

DOI: 10.4324/9781003274414-1

1   Introduction to UX Writing

Chapter Overview

This chapter offers an introduction to the characteristics of UX writing. It defines the scope and emerging practices of UX writing for readers new to UX as well as those who have worked in traditional technical communication areas like documentation writing, information design, content strategy and management, copywriting, and editing.

Learning Objectives

- Understand and define writing for user experience.
- Articulate the relationship between UX writing and adjacent professions.
- Distinguish the needs for UX writing in designing products and services.
- Outline the motivations to study UX writing.

A Brave New World

Imagine that you are a newly hired writer at Microcorp who has been put in a team that is working on the next release of Operating System Z (Figure 1.1). You have not met with the entire team, but you have already been asked to give a brief presentation on how the design team should begin working on the release. You may be asking yourself: What does a *writer* have to offer to the *design* of a product? Why put a writer on the team in the first place? What would you do in that scenario? If you were a technical writer in the 1990s, you would probably wait for directives from product engineers to write documentation after the prototype was done. But those days of using documentation as a "band-aid" for mediocre user interfaces are long gone. In the modern workplace, you will collaborate with the engineers in designing the experience for your users from the get-go. And this is because organizations and companies know nowadays that the user experience is paramount. Simply put, if a user isn't happy with their initial experience with a service, they will most likely not continue to use it.

The role of a writer has evolved to be an integral part of the design process. According to design strategist Leah Buley (2013), nowadays many roles in the tech industry are expected to be a "team of one" and be able to fulfill many different roles in organizations, and this is particularly the case with writers. Writers contribute directly to the design of a meaningful

DOI: 10.4324/9781003274414-2

Figure 1.1 A UX writer presenting a content strategy to a design team.

Source: Jason Goodman on Unsplash.

experience between the end user and a product or service. In this opening chapter, you will learn about the changing characteristics of writers in the digital age and the emerging practice of *user experience writing* or UX writing.

UX: The Design of Experience

To understand user experience writing, we need to begin by defining user experience. Often shortened as UX, user experience "encompasses all aspects of the end-user's interaction with the company, its services, and its products" (Norman & Nielsen, 2021). To put it simply, UX is concerned with the human experience when interacting with an interface like a website, user manual, or instructional video. Depending on the context of *use*, a user can be an actual or potential reader, viewer, shopper, customer, or consumer. In the digital age, users are afforded access to information and services from all around the world through the internet and the web. While in this chapter and throughout the book we commonly refer to UX examples that are online or digital, you should also apply the same concepts to local and physical situations where user interactions happen every day.

Marketing professionals already know that good UX is important to any product or service because we have to incentivize and attract users to a product by giving them something they want or need. Herbert A. Simon called this the "attention economy" (1971, pp. 37–52). We have to "remunerate" consumers' attention, and the "economy" here is pretty straightforward and commonsensical. Users have a limited amount of time and capital to invest in their online experiences. Plus, they have a lot of competition placing demands on that temporal capital, or, as Mitch Kapor, a pioneer of the personal computing industry and long-time startup investor, put it, we are always in "a competition for eyeballs" in the marketplace (Malik, 2003, n.p.). As writers, therefore, our job is to figure out how to convince users that the time they spend using our product is the best investment of their limited time and capital. For most of us, the answer to this question is going to involve providing a richer, more satisfying user experience.

We are essentially looking for ways to attract users to our products by convincing them that the time they invest in us will be rewarded with the experience they seek. The analogy that we think works well here is that of going to a restaurant. In his book, *Design to Thrive*, Tharon wrote:

> If you think about it, you don't really go to a restaurant to fulfill your body's needs for sustenance. You could go to the grocery store, buy bread and peanut butter, and make yourself a sandwich if all you wanted to do is feed yourself, right? You're really after something more than just food when you go to an expensive restaurant for dinner. What, for example, is the attraction of Japanese steak houses and sitting around a hibachi grill with a bunch of other people—usually complete strangers—watching your food cooked in front of you and worrying if your hair is going to be singed by the obligatory grease fire that you know is coming? Indeed, I asked my 18-year-old son about this recently when he chose to have the family go to a Japanese steak house in order to celebrate his birthday. Wouldn't you, I asked him, rather go to the local, family owned steakhouse which serves a better quality 20-oz. Angus Porterhouse for less money than the 5-oz beef tips and 6 small shrimp you're going to get at a hibachi grill? What's the attraction, I asked. Wouldn't you rather have a better steak?
>
> Of course, answering a question like that takes far too much conscious effort on the part of an 18-year-old, so I got the same, long suffering, "you're-so-clueless-Dad" look that I get when I ask questions about what makes being a member of a *World of Warcraft* guild so compelling. Still, I think the answer is pretty obvious. It's not only about the food; you're also paying for the atmosphere, the service, and most of all, it's about the entertainment provided by the chef. The remuneration is *the experience*.
>
> (Howard, 2010, p. 44)

So at this point, the importance of providing users, customers, and audiences with a positive user experience probably seems pretty commonsensical. Unfortunately, the problem is that it is *so* commonsensical that most people (especially writers and designers) never think about it. We have a tendency to take good UX for granted. As renowned architect and system theorist Buckminster Fuller famously said, "Ninety-nine percent of who you are is invisible and untouchable" (R. Buckminster Fuller Quotes, n.d., n.p.). In other words, we're so immersed in the business of completing tasks and getting on with our lives that we're usually no more conscious of user experiences than a fish is conscious of water. Good UX is part of the air we breathe, which is why it's nearly invisible—just like the popular design podcast by Roman Mars also titled "99% Invisible" (https://99percentinvisible.org/).

While a person's experience with a product may be so tacit that it's unconscious and invisible, the key takeaway here is to remember that good UX doesn't just happen. Good UX is *designed* and *intentional*, even though users don't always notice it. How often, for example, do you think about the products and components in your bed when you lay down to sleep at night? If you're like most people, chances are you snuggle under the sheets and blankets and curl up on your pillow without a conscious thought about *why* the experience feels good. It just *feels*. And yet every aspect of that experience is the result of very

Figure 1.2 Grease fire entertainment by a hibachi grill chef.
Source: Howard (2010).

careful and conscious design decisions by a large number of product design teams. Take the bedsheets, for example. Most people won't be able to tell you the "thread count" of the sheets that they use every night, but the designers of your sheets certainly can. They know that thread count is a measure of the number of horizontal and vertical threads per square inch in the cloth, and (more importantly) the higher the thread count, the softer the sheet and the more likely it will be to soften over time. Good sheets have thread counts ranging anywhere from 200 to 800, and the highest-quality sheets have over 1000. And thread count is only one factor in the UX design for a sheet; designers spend extraordinary amounts of time researching materials (cotton, linen, bamboo, etc.), weave patterns (percale weave, sateen weave, etc.), and new manufacturing processes. Sheet designers invest tremendous amounts of thought in finding the best balance between the cost and the user experience of our bed sheets so that we don't have to. And the same thing goes for your pillow, mattress, bed springs, blanket, bedframe, quilts, etc.

However, while good UX design is often so tacit that it's unconscious and invisible to the end user, the same can't be said for poor UX. Earlier, we said that we are often no more conscious of user experiences than a fish is conscious of water. But we definitely notice bad UX design. When you try to make a purchase on a shopping website, but the interface won't allow you to select the item you wish to buy, that's bad UX—and your frustration makes you notice it. And you probably won't come back to the website if it is frustrating.

One really famous example of people noticing bad UX is from former Apple employees, Donald Norman and Bruce Tognazzini. Tognazzini (better known as "Tog") was Apple's very first application software engineer, and he is famous because his early usability testing and work with software interfaces in the 1970s led to Tog's "Apple Human Interface Guidelines," which provided the foundation for Apple's early success in the 1990s. Through Norman and Tog's work, Apple became successful because of the simplicity and ease of use of its operating systems and software application designs. However, this all changed when Apple developed its gestural user interface for use with iPhones and touchscreen devices operated through hand movements and gestures. Apple abandoned the user interface guidelines that had made the Apple II and the Macintosh successful, and they adopted an approach that, according to Norman and Tog, focused on "look" rather than "feel." In their critique of the new gestural interface designs, Norman and Tog famously wrote:

> Apple is destroying design. Worse, it is revitalizing the old belief that design is only about making things look pretty. No, not so! Design is a way of thinking, of determining people's true, underlying needs, and then delivering products and services that help them. Design combines an understanding of people, technology, society, and business. The production of beautiful objects is only one small component of modern design: Designers today work on such problems as the design of cities, of transportation systems, of health care. Apple is reinforcing the old, discredited idea that the designer's sole job is to make things beautiful, even at the expense of providing the right functions, aiding understandability, and ensuring ease of use.
>
> (Norman & Tognazzini, 2015, np)

One of the things that makes this quote so famous is that it is full of lessons for students of UX and design. Of course, the first of these is that users *notice* and respond to bad UX. But the quote also provides one of the best and most succinct definitions of modern UX design. It rejects the idea that designers merely make things look nice and pretty. We tend to think that, for example, an interior designer's job is primarily to pick out complimentary colors for furniture and wallpaper, but the definition here extends the idea that *design is intentional* because it recognizes that it "combines an understanding of people, technology, society, and business." In other words, it requires research into what users need in all aspects of their lives. It begins by defining users' problems and the environments in which those problems emerge and then designs experiences that solve those problems.

And even though it looks invisible and easy, good UX design is actually really hard. And because UX design is challenging and in high demand by organizations, some jokingly refer to the desire of companies to hire a "UX Unicorn" who can do it all, combine a great deal of skill sets, and do all of these perfectly. That is why all three of us have an image of the UX unicorn on our office door, to remind our students that you cannot expect to do it all perfectly but that you do need to know something about each of these areas.

The Rise of UX Writing

In an age of content marketing and digital design—where users actively seek information and where marketers find it an opportunity to sell products through information services and experience design—content-first design promises short-term as well as long-term success. UX-centric content makes for efficient design and a greater return on investment (King, 2022). The idea of content-first design is based on figuring out what content your user is looking for before designing the interface, rather than creating an interface and hoping to fill it with relevant content (Johnson, 2020). The same principle applies to UX writers—they have to know how good *content* can be created first so the *design meets the user's immediate needs and wants.*

Many companies hire professionals trained in UX to support their product and service designs. For instance, "UX researcher" and "UX designer" are well-established career positions that involve the study of consumer experience and the application of insights to the design of products and services. A UX researcher develops and conducts user research such as UX journey mapping, persona development, ethnographic studies, or think-aloud protocol analyses, whereas a UX designer's role is more strategic—devising product structure, content strategy plan, and prototypes using the data collected by the UX researchers. In some companies, these roles converge and could be performed by a UX specialist or a UX team.

We are seeing an increase in the number of job descriptions that appeal to writers who care about UX (see Snapshot 1.1). Many companies are seeking to hire writers with UX expertise to work in collaboration with designers to integrate content with interface design in order to cultivate a good user experience.

Real World Snapshot 1.1: A UX Writing Job Description (Archived in 2021)

UX Content Writer – Opportunity for Working Remotely!
Location: Boulder, CO – Remote
Salary: $95,000 – $162,000 a year
Job Type: Full time
Full Job Description:
We are looking for a UX Content Writer with experience in complex cloud applications to help us transform the cybersecurity industry. In this role, you'll collaborate with cross-functional

teams to design and deliver solutions that make it easy for our customers to make their organizations more secure. You'll use your UX knowledge, experience, and judgment to help us move away from traditional manuals, designing and integrating lightweight, targeted assistance directly into the platform UI. Your UX content design skills are integral to the continued success of our industry-leading cybersecurity platform.

Since our experience with products and services is heavily influenced by the content we consume—think about the number of posts, text messages, videos, memes, and audio signals you encounter each day—UX writing is concerned with the integrative experience between the user and product/service as it is mediated by different content. They include texts or copy, images, videos/animations, sounds, haptics (touch senses), or a combination of them. Good UX writing leads to a seamlessly positive experience between the user and the product.

UX writing isn't just a product, however. UX writing is also a process involving writing, designing, thinking, iterating, strategizing, and developing content using different technologies to create a desirable user experience. This integrative approach urges you to see content development as an interconnected activity that leads you down many different paths with one goal: making a better user experience for the user.

Now, think about your smartphone (well, if you use one, or else ask a friend who does). What was your experience like when you took it out of the box for the first time? Looking at the shiny black mirror that reflects your excitement, what was your initial reaction? Were you confused by it? Did you feel the need to read the user manual? Like most users, we bet you just turned on your new phone and started swiping away immediately. You're greeted with a friendly welcome message and a quick-start wizard that walks you through the setup process. And if it so happened that you stumbled upon something you were unsure of during the setup, like pairing the phone with other devices you own, you'd have probably quickly found the option to just skip it. Just like that, in less than a few minutes, you were all set and ready to post a new selfie on your Instagram with a whimsy hashtag like *#newphonewhothis*.

People want to enjoy products rather than learn how to use them through user manuals. UX writing eliminates the dreadful process that hinders good interactions by designing content that facilitates a desirable user experience (Figure 1.3).

Writing at Multiple Intersections

Good UX writing helps users perform desired tasks while serving the needs of marketing or selling a product/service. To be successful in UX writing, you need to develop expertise at the intersection of several connected areas, such as user experience research, content strategy, and interface design, to list a few. The UX writing compass (Figure 1.4) is our attempt to map out the different expertise required for doing UX writing. At times, your job may require you to focus on one area, and other times you may need to draw from each of the directions to consider how to develop your content for users more successfully. The compass helps you consider which areas you need to draw upon.

UX writers need to develop a 360-degree situational awareness so they know which competencies *and* complementary skills they need to draw from (e.g., design thinking, content

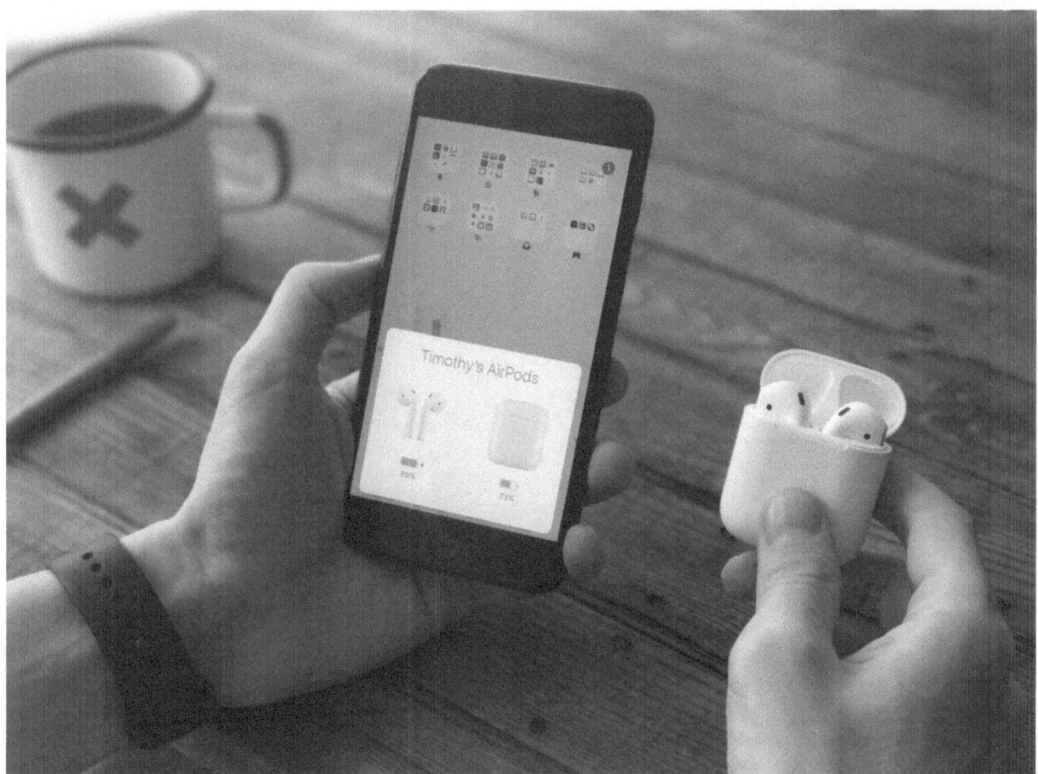

Figure 1.3 UX writing creates positive, seamless product experiences.

Source: Timothy Buck on Unsplash.

strategy, and user experience, among others). For instance, they may be involved in developing a content strategy plan at work but realize that, per content-first design principles, they may first need to do design thinking to help formulate a marketing strategy with a content team. This may lead to a discussion about the user experience of content before a content strategy can be developed. Hence, having a compass to consider where you are is important, as is knowing where you need to head next.

It is important to keep in mind that UX is always-already interdisciplinary. The relationship between technical communication and UX has been seen as "intertwined" throughout the last few decades (Redish, 2010; Redish & Barnum, 2011). Many UX practitioners today come from the line of technical writers who were initially trained to perform conventional writing tasks like documentation, technical editing, and publishing. Since the 1980s, however, the growth in information technology and personal computing has shifted the role of technical communicators into that of information designers and content specialists. The increase in desktop publishing tools and user-generated content forces writers to not only be information producers but also experts in identifying, curating, and repurposing good content to serve organizational needs.

Moreover, writers are expected to conduct field research and user studies in order to gauge customer experience and needs. The popularization of design thinking and user-centered design in the early 1990s has further expanded the role of the writer in product

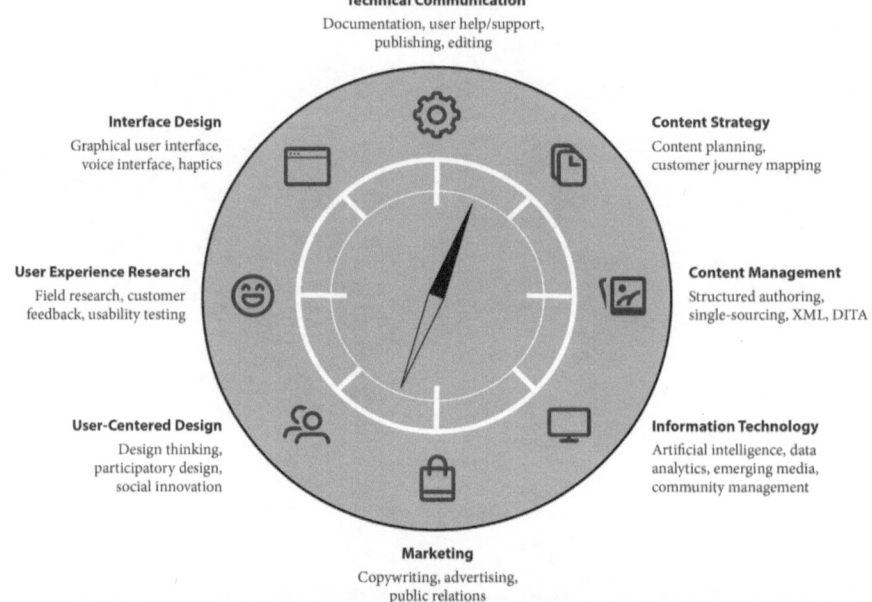

Figure 1.4 The UX writing compass.

development. Writers have now become an integral part of the ideation and innovation processes in product design; they participate significantly in the processes of generating user personas, creating product prototypes, and conducting usability tests. They then present recommendations to stakeholders and designers to guide future development.

As digital products and services become more and more ubiquitous and pervasive in our everyday lives, writers become increasingly involved in the planning, deployment, and management of content on our screens and beyond. UX writers learn to map out a customer's "journey" when interfacing with a product or service, compose reusable (or modular) content through single-sourcing, craft materials that can be accessed by multiple modes of delivery (voice, sounds, words, and visuals), and engage with new and emerging technologies such as social media and artificial intelligence (AI) to craft relevant and desirable content. Companies that provide digital services want to ensure an optimal user experience through impactful content. In many cases, writers who are now content designers focus on "answering a user's need in the best possible way for the user to consume it" (Winters, 2019, n.p.). This is especially important when content is now accessed using different channels, including webpages, mobile devices, voice interfaces, and streaming services, among others.

Content strategists research how users typically access content and strategize how content should be delivered from a company or business to a user. As companies seek to deliver engaging content, writers focus on "business goals, organizational context, and user needs, and then which tactical activities are necessary to achieve this success" (Batova & Andersen,

Figure 1.5 Writers perform many roles in designing the user experience.

2016, p. 2). Content strategists thus think about the marketing of content in relation to the goals of the organization and how consistent it appears to the user.

Most often, companies hire UX writers to do all of the above tasks, which is why UX writing is growing as a popular profession. In this book, we will use the term "UX writer" to encompass all these different task areas rather than focusing on a specific role in writing. We see UX writing in the notion of "one job, many hats"—that UX writers perform some or more roles in planning, researching, strategizing, ideating, prototyping, and testing while on the job (see Figure 1.5).

Real World Snapshot 1.2: Vignettes of UX Writers' Work Environments: Larry's Job

Larry works for a Fortune 100 company that, for more than 90 years, has been one of the leading distributors of maintenance, repair, and operating products and services to U.S. companies and government agencies. As a business-to-business service provider, they have a wide assortment of services, deep expertise, and innovative technology solutions designed to keep their customers' operations running and their employees safe. If, for example, a company has a large fleet of delivery trucks that need to be maintained, Larry's company offers the tools needed to handle everything from when to service engines, pay insurance, rotate drivers' schedules, and/or purchase replacement vehicles. And even though it's been around for almost nine decades, the company's success has come from embracing technology and change and doing deep, introspective work to deliver the best results for their customers and field teams. The company's mantra is "start with the customer."

Larry's title is Senior UX Manager, and the tool Larry works on is called "FixStock." FixStock is a $1 billion inventory management service that runs on a variety of computer operating systems and mobile devices and helps companies track when they need to replenish stock, rotate and refresh expired items, maintain storage facilities (like freezers and vending machines), and much, much more. FixStock has 1,800 field team members who support the software and services, and as FixStock's Senior UX Manager for the Midwest, Larry's job is to evaluate and recommend ongoing improvements to make the tool the "best-in-class" service for both the company's customers and the 1,800 field team members who support them. His job is to continue the company's "start with the customer" mantra by integrating research into their work and by measuring the usability of their web and mobile experiences.

As the Senior UX Manager's job description states, some of Larry's key tasks include:

- Providing expertise in planning and conducting exploratory, generative, and evaluative design research.
- Collaborating with UX Design, Product Management, Analytics, and Business Partners to translate business needs into research questions and hypotheses.
- Arming product teams with actionable insights to inform the user experience and product roadmaps.
- Owning a research roadmap that will create transparency across product teams and ensure insights are delivered when needed by product teams.
- Increasing empathy across the organization by establishing a research program that will give product teams and business partners more exposure to the attitudes and behaviors of the company's users.
- Helping build and manage a customer research panel that will enable us to plan and run studies more efficiently.
- Helping formalize and operationalize UX Metrics.
- Determining what research tools we need to effectively plan and conduct research.
- Managing and growing our UX research and measurement talent.

As Larry's job description shows, his position is both complex and interesting, and it demands far more writing skills than simple wordsmithing, copyediting, and "microcopy" content creation.

Note: The Real World Snapshots throughout this book are based on real people at real companies; however, the names have been changed to protect their confidentiality.

UX Writing Goals

One important distinction between UX writing and the conventional understanding of technical and professional writing (writing on the job) is that UX writing goes beyond content creation. We will show a few examples later on. But it is important for those who perform UX writing to understand that UX writing is writing *within* and *around* an interface. UX writers think like product/service designers and use content to create an optimal user experience. Below, we feature a few key characteristics of UX writing that we'll be covering more in-depth in the rest of this book.

Writing to Gather Attention. We earlier referred to the smartphone scenario about how UX writing is imperative to the whole "out-of-the-box" user experience. Now picture companies seeking to understand what users do when they interact with content on their devices as they go about their day. UX writers not only write the content for such exchanges; they are also tasked with keeping track of what users say, think, and feel during these exchanges as they interact with products or services. And as companies seek to capture the user's attention, they need writers who understand distracted users and who can develop personalized content specific to individual users. For instance, you may want to know about local restaurants that you can visit while driving to your destination using your phone's map or global positioning system (GPS). Companies understand that having a website is not enough to attract these users, so they are looking to UX writers who understand users and can develop content for situations like this.

Writing Intelligent Content. A seismic shift has taken place: content is no longer created on the basis of users accessing a website. Instead, *intelligent* content (such as restaurant recommendations based on your location or based on similar types of restaurants you've ranked highly in the past) is now expected by users. Users want content that is easily accessible, specific, and personalized for their circumstances rather than something they have to actively look for on a website. Content should be able to be retrieved by users for a wide variety of situations, but it also needs to be specific to what the user needs. And content needs to be read by machines so a search engine can recommend restaurants near your GPS signal. Indeed, companies are now "separating [their] ... interface designs from the underlying content structure" to address this new expectation for content (Atherton & Hane, 2018, p. 12). UX writers are important in these situations because they not only write content but also understand content strategies to help push intelligent content to users.

Writing Sharable Content. As we know, users may now share content through various digital channels (social media, web, apps, etc.) while using their mobile device while out in the world. For instance, think about how you can modify your plans while you are meeting up with a friend by sending them a new location through social media. Others may comment, like, or share that content with others in your group to let them know where you are meeting now. In other words, content is now *social*. Now picture companies knowing this, and you will understand why companies now heavily invest in user experience design. They want to ensure their product, service, or design works well for many users and avoid negative reviews from groups of people.

Writing to Influence User Behaviors. In order to ensure that users keep using products and services, companies now also use what are called "nudges" to help influence the user's behavior. A "nudge" is a way of designing the interface to make users want to come back or perform a certain action. Companies know that users remember extremely positive and negative experiences. This has led to the adoption of design techniques that seek to alter how users respond to experiences by motivating them to perform a certain action. For instance, Amazon.com uses many "nudges," such as showing how many people bought a product and featuring reviews, so that people are ensured to buy the product with confidence when they visit a product page. And when a search error occurs (a negative experience), Google displays a funny image of a robot so as to ensure the user is not alarmed and has a positive experience instead of a negative one (Figure 1.6).

Writing to Gather Data. Content and UX design thus now form the basis for companies to engage users through ongoing experiences in various channels (social media, web, apps, and desktop) as they go about their day. For this reason, companies keep track of content *before, during,* and *after* an interaction. As people now spend a large amount of time online, users' behavior can be seen by what they do while they access content online. Do they spend a lot of time on your page or your app, or do they quickly navigate away? If they spend a lot

1. Error notification.

2. Image of robot nudges users to remember the silly robot with broken parts, and so using the peak-end bias.

Figure 1.6 UX writing uses techniques to ensure the user responds in a positive way to the error message by focusing instead on the funny-looking robot.

Source: Verhulsdonck and Shalamova (2020).

of time there, does that mean the content you offer is successful and lets them do what they need to, or are they spending the time there because they're confused by the content? UX writers thus may also need to research what users are doing as they interact with content. Various user metrics exist to understand what users do and think, such as time-on-task and user evaluation scales. As a result, companies hire writers who can address these different skills through writing, research, and content design.

Writing to Enhance Usability. On a finer level, UX writers pay attention to details that affect the user's interaction with a product, such as microcopy content on an interface. In Figure 1.7, you can find a comparison between two simple microcopy designs. Microcopy refers to short texts on websites, digital applications, and interactive interfaces that provide information or instructions to users. More examples of these genres can be found in Chapter 9.

In this example, the two forms contain almost exactly the same information for a user to fill out the fields; however, due to the positioning of error messages on the left form, users may be confused by the errors and would need more context to correct the errors. In comparison, the microcopy design on the right places the error message and relevant corrective options exactly where the actions need to be taken to help the user quickly recover from the error. As we can see from this instance, it is a writer's job to recognize the best practices in such designs as microcopy to ensure a positive user experience as well as usability and efficiency.

Writing to Sell. UX writing isn't just about interface or content design. As we have indicated in the UX writing compass, part of the responsibility of UX writing is to ensure that whatever content (written or otherwise) is set to appear before a user or consumer should work to achieve marketing objectives. In other words, writers work closely with marketing and sales teams to create content that entices users by making the product or service desirable.

In sum, UX writing bridges technical writing and creative copywriting, which are traditionally absent from technical communication training. A UX writer needs the skill to

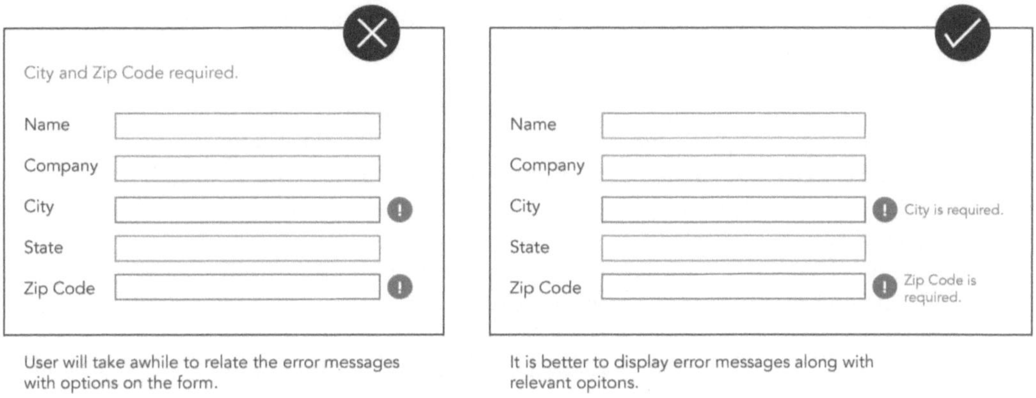

Figure 1.7 UX writing is writing to enhance usability via design.

poetically weave technical specifications and product information with strategic content that promotes sales and attracts customers. This can only be done if the writer is mindful of the user's needs and experience.

UX Writing Technologies: AI, Data Analytics, Oh My!

In the example we noted earlier, UX writers produce *intelligent content* that lets users find restaurants near where they are going using just their mobile device. How is that done? UX writing uses a number of technologies to make sure personalized content reaches the intended audience. As you know, GPS signals indicate where you are, and that data can be used by a search engine to find restaurants near you. But they have to be good restaurants and ones you are interested in. For instance, you may want to go for Mexican food when going to an unfamiliar destination before meeting your friend. You may use voice input because you are driving and ask Google for "Mexican restaurants near me," and you expect your mobile phone to notify you about relevant restaurants with good reviews. It even offers to modify your driving route to go to the restaurant before you meet with your friend.

Companies understand that users come to them with problems to solve or tasks to accomplish, and they need writers to clarify this process for users. In this case, UX writers understand that the user is *really* requesting, "Look for *good* Mexican restaurants near my GPS location and tell me about them." A number of technologies are working together to produce this effect:

- **Locative technologies** (like GPS): to locate where you are when you ask this question.
- **Voice recognition:** to translate the words you speak ("Mexican restaurants near me").
- **Natural language processing:** to translate what you are saying into a command and to analyze what it means ("Look for *good* Mexican restaurants near me").
- **Data analytics:** to analyze your question and match it with an answer based on online information ("Check starred reviews online on good Mexican restaurants and aggregate them for me, and tell me the best ones").
- **AI:** to formulate an intelligible response using natural speech that you can hear as you drive ("Here are some recommendations for good Mexican restaurants near you").

And this is just one instance of technology's use. Depending on the context, UX writers select appropriate technologies to plan, create, organize, store, and publish their content:

- Cloud-based drives for document sharing and collaborative composing.
- Graphic design suites for technical illustration and document design.
- Component content management systems for modular editing and publishing.
- Markup languages and standards for structured authoring.
- Data visualization tools for complex information.
- Prototyping and modeling applications for interaction design.
- Social media community management platforms.

All of this is to say that UX writers know that a mix of good content, understanding the user and their task, and how various technologies work can help create powerful experiences for users that will, in turn, add value to companies and organizations. And this is why UX writers can expect to do more content design in the future: as companies use various technologies to deliver content, users evolve alongside them and demand dynamic solutions to their tasks through intelligent content.

Real World Snapshot 1.3: Vignettes of UX Writers' Work Environments: Laurie's Job

While Larry is an example of someone who works in a department for a large, multi-site corporation, another common situation in which UX writers find themselves is exemplified by Laurie. Many companies and government organizations choose not to maintain internal departments and staffs of specialists. Instead, they outsource their application development needs.

Because they may not need new software applications for their operations once every three to five years, it's simply not cost-effective for them to pay for the salaries, insurance, retirement benefits, and other overhead costs associated with maintaining full-time employees on their permanent staff. Instead, they hire companies that specialize in providing software and web application services on a contractual basis. Laurie works for one of these, and her company, Namaste Labs, specializes in projects for federal and state agencies as well as large corporations like Walmart, Lowes, and Magnavox.

Namaste Labs needs to meet their clients' specific needs in a variety of computing environments (e.g., mobile device applications, websites, mainframe databases, and iOS). Also, they need to understand their clients' task environments (what goals do users have for tasks, where do they complete the tasks, what steps are involved, what are the successful completion criteria for the tasks, etc.). They also need to develop graphics for the interfaces, microcopy for the pull-down menus and other text in the interface, context-sensitive help systems, training videos, and much more. Since clients' needs will vary considerably from situation to situation, Namaste Labs has project managers who work with writers to create proposals specifically tailored to meet each client's situation, and for each case, they will create a cross-functional development team comprised of UX researchers, UX writers, graphic designers, UI interaction designers, project managers, and/or engineers necessary to meet the needs of the project.

Like Larry, Laurie's role as a UX writer is rarely limited to simply writing the microcopy for an interface. She is often involved before the team is even created because she works with the project manager on writing the proposal for clients and even deciding which specialists at Namaste Labs will be needed for the team. She's also involved in conducting research on the end users of a product being created or updated for a project and is responsible for conducting interviews and site visits with end users and conducting task analyses of their work. Like Larry, she's involved in creating personas of different user types from the demographic data she collect, creating user experience maps that allow the whole project team to visualize each "touchpoint" where users interact with a product so that the team can design for those experiences, and much, much more. Laurie's knowledge of the users means she's the member of the team who creates the microcopy for the UI since she knows the terms users are seeking when they are looking for pulldown menus or buttons to complete a task. She's able to create effective instructions on how to complete specific tasks, and she's able to work with the graphic designer to develop visual metaphors and icons that allow users to utilize the interface more successfully. And because of her task analysis research, she helps the software engineers break the tasks users need to perform down into manageable parts so that the engineers can write their code in modules. Bottom line: every team Laurie supports has different needs and different members, so (like Larry), she needs to bring a far wider range of skills to her job than her job title, UX Writer, might suggest.

Note: The Real World Snapshots throughout this book are based on real people at real companies; however, the names have been changed to protect their confidentiality.

UX Writing Career Facts

Even though you may never have heard of the term "UX writer" before you picked up this book, the fact is that many professionals have been and are pursuing careers with the job title UX writer. And even if you knew that you could find jobs with the title UX writer, many people are surprised to learn that some of the highest-paying jobs for people who write in the tech industry aren't held by employees with the title "technical writer" or even "information product developer." Instead, they're actually held by people whose job title is "UX writer" or "content designer." Table 1.1 is from a 2021 survey by Yuval Keshtcher of almost 800 writers around the world who work in the high-tech industry. According to the Bureau of Labor Statistics' (BLS) Occupational Outlook Handbook, the median salary of technical writers in the United States during 2021 was $74,650 per year (BLS, 2021); however, as Keshtcher's table shows, it's possible for UX writers to earn far, far more.

Keshtcher's survey not only showed that the highest-paying positions were held by people with the UX writer job title, but he also found that the median salary was higher. He wrote, "You often hear that the job title doesn't matter. Well, in terms of salary, that's not true. There is a clear tendency for some job titles to earn more than others" (n.p.). Here are the median salaries for these common job titles:

As you can see, the top three high-earning titles are content designers, UX writers, and content strategists (Keshtcher, 2021, n.p.).

In a different 2019 survey of over 200 UX writers around the world, Patrick Stafford (2019) from the UX Writers Collective found that:

- 26% of respondents were identified as male across all job titles.
- 74% of respondents were identified as female or "mostly female."
- 31% claimed 10+ years of experience.
- 44% claimed between 3 and 9 years of experience.
- 21% claimed between 1 and 3 years of experience
- 4% claimed less than 1 year of experience

Evidently, the field of UX writing is growing. It is a fine career for many writers who are interested in creating desirable customer and user experiences for any product. Our book is designed to give you tips and guidance on developing skills that can lead to a promising career in UX writing. In the following chapters, you will learn about principles and concepts for effective UX writing, methods and technologies that are used in UX writing, and projects that let you exercise UX writing and produce your own UX writing portfolio.

Table 1.1 Highest-paying salaries for writers in tech around the world by job title (Keshtcher, 2021)

Country	Max reported annual salary	Job title
United Kingdom	$376,895	Content Designer
USA	$360,000	Content Designer
Sweden	$240,479	UX Writer
Germany	$194,085	UX Copywriter
Switzerland	$168,527	Content Strategist
Netherlands	$163,759	Content Strategist
Canada	$146,000	Content Designer
Ireland	$133,640	UX Writer
Australia	$116,912	UX Writer

Table 1.2 The median salary for UX writing job titles in the US.

Job Title	Median annual salary (USA)
Content Designer	$117,500
UX Writer	$110,000
Content Strategist	$105,600
Conversation Designer	$105,000
Content Director	$100,000
Technical Writer	$93,000
Content Marketer	$80,000
Content Writer	$70,800
Content/Copy Editor	$70,000
Copywriter	$65,449

Conclusion

UX writing is becoming increasingly important to our professional and everyday lives. As you learned in this opening chapter, good UX is *designed*, and it is *intentional* in bringing us good experiences with products and services. Yet, because good UX is invisible but very important to how we feel about our experiences (which will determine whether we continue with a product or service), UX writing is gaining prominence in the technical and service industries. Due to the limited amount of attention of users with requirements for pleasant, memorable, and useful experiences, companies are now looking for writers who combine strong UX writing skills with competencies in copywriting, content design, and data analysis to create usable content as well as user interfaces.

Rather than merely writing successful content for an interface, writers need expertise in diverse areas to address how content functions *in*—and equally important, *around*—the interface. This is because UX writing is about creating successful content for products and services as experiences for different users across numerous channels. Users want a consistent experience with a product or service from a company. Yet companies know that such experiences happen across various online and physical channels, which require strategizing to provide a consistent experience for their users. As UX writing isn't just about transmitting content but also about adapting the experiences of users as they evolve, this requires different skills beyond traditional technical or professional writing. Due to the need to develop memorable content, UX writing not only bridges technical writing skills but also bridges creative copywriting, product development, and content strategy to promote sales and attract customers.

In thinking about how UX writing applies to your life, consider how you continue to use various products and services in your everyday life. As we have shown as examples in this chapter, your phone lets you send messages, use various apps, and connect with your friends and family in an easy and pleasant manner. A restaurant you frequent offers not only good food and ambience but also attentive staff and an entertaining experience. A ride-share service such as Uber lets you quickly and easily order a ride and travel to places without the hassle of waiting and paying cash. All of these instances involve content as part of a user experience that needs to be pleasant, memorable, and easy to use. Now think of those products you are not using anymore. Most likely, they failed to provide a pleasant experience, which shows how incredibly important UX is in a product or service. UX writing delivers products and services in intentional and designed ways so you will continue to use them.

Chapter Checklist

- *Good UX is about designing for experiences that are pleasant, memorable, and useful.*

  - UX is concerned with the human experience when interacting with an interface.
  - UX design is intentional.

- *Successful UX is invisible to the user but will determine if they will continue with a product or service.*

  - Good UX design is often so tacit that it's unconscious and invisible to the end user; the same can't be said for poor UX.

- *UX writing is important to business.*

  - Companies are looking to engage users with limited attention who want positive experiences using different products and services.
  - UX-centric content makes for efficient design and a greater return on investment.

- *UX writing is everywhere.*

  - Content is accessed through different channels, including webpages, mobile devices, voice interfaces, and streaming services, among others.

- *UX writing is writing in and around the interface and focuses on before, during, and after a user interacts with content.*

  - UX writing isn't just a product; it's also a process involving writing, designing, thinking, iterating, strategizing, and developing content using different technologies to create a better user experience.

- *UX writing is multifaceted.*

  - UX writing not only combines technical writing skills but also combines creative copywriting, product development, and content strategy to promote sales and attract customers.
  - UX writers need to develop a 360-degree situational awareness so they know which competencies *and* complementary skills they need to draw from.
  - UX writing involves writing to gather the user's attention, creating a strong and seamless product experience, developing intelligent content to push to the user, gathering data on your users, and aligning content with marketing goals for your users.

Discussion Questions

1 You've probably worked in a team before, maybe on a group assignment or on a committee for your student organization. How does writing happen in your team? Who determines the quality of the content? How do decisions get made?
2 Observe any written content or documents around your workspace (like an employee handbook or weekly schedule) or online (social media, emails, or your school's website). Which content is visible and which content is "invisible" on these documents? How do you consider that distinction?
3 Recall a bad/unpleasant experience that you had with a product or service recently (e.g., using your school's learning management system, trying to understand an assignment rubric, resetting the clock on your microwave). What made that particular experience a bad experience? How did you wish the experience would have turned out differently?
4 Thinking about the same bad experience you had, consider what kinds of content could have been improved to mitigate the bad experience. If you were to rewrite them, what would you change?

Learning Activity 1

You have been asked to join a user survey group for your school's student portal redesign efforts. You are a sophomore who has used the student portal regularly to register for classes, see tuition balances, download forms for your student organization, and check your final grades. Now, pretend that the following screenshot represents the current version of the student portal. You agree that it has been an awful experience navigating the portal. In the user survey group, the research facilitators have asked you to sketch out your vision for a better student portal. Keeping the needs you have in mind, sketch out a redesigned version of the portal that would best serve your needs. Don't worry about keeping any existing

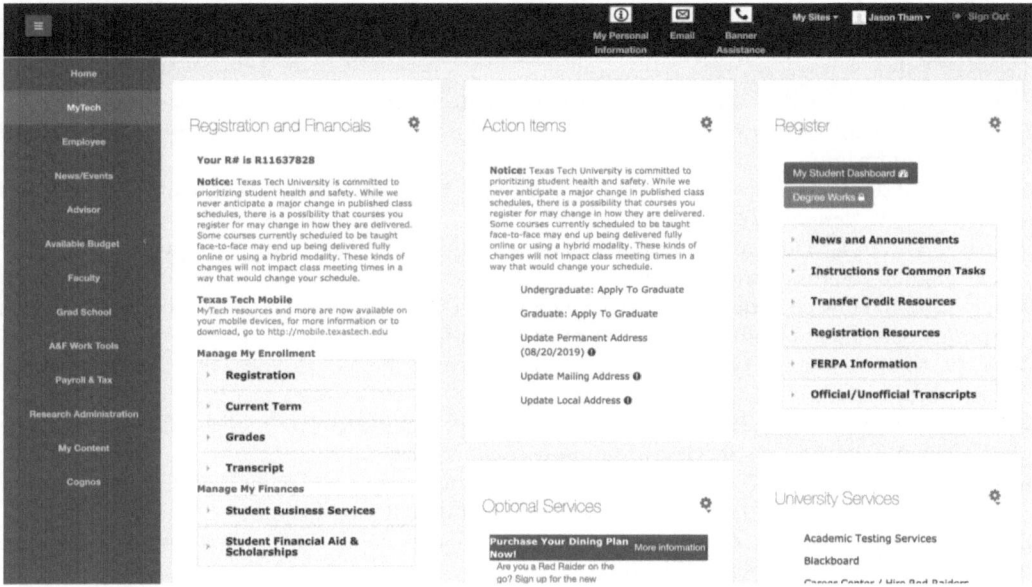

Figure 1.8 A current student portal.

Source: Author's screenshot.

content. Don't worry about making the sketch perfect. Just pull out a piece of paper and start pencil-sketching your idea (Figure 1.8).

Learning Activity 2

UX writing will continue to grow as a profession in technical and service industries since we and our world are continuously becoming more reliant on good content to help us do our work and improve our lives. Many technical communicators and professional writers are finding themselves in UX writing roles; they are contributing to the formative identity of a UX writer at a time when design and writing become increasingly intertwined and integrated in product and content development.

1 Perform a web search with the keywords "UX," "writing," "content," "design," and others that may be relevant to your career aspirations. Refer to the UX writing compass back in Chapter 1 (Figure 1.5) for further inspiration.
2 Summarize your search results in a Venn diagram. What are the main domains that UX writing occupies? What are the overlapping traits or practices?
3 Share your Venn diagram with a friend or classmate. Tell them where you see yourself in the picture.

References

Atherton, M. & Hane, C. (2018). *Designing connected content: Plan and model digital products for today and tomorrow.* New Riders.

Batova, T. & Andersen, R. (2016). Introduction to the special issue: Content strategy–a unifying vision. *IEEE Transactions on Professional Communication, 59*(1), 2–6. https://ieeexplore.ieee.org/document/7448985

Buley, L. (2013). *The user experience team of one: A research and design survival guide.* Rosenfeld Media.

Bureau of Labor Statistics (2021). Technical writer. *Occupational outlook handbook.* https://www.bls.gov/ooh/media-and-communication/technical-writers.htm

Howard, T. (2010). *Design to thrive: Creating social networks and online communities that last.* Morgan Kaufmann.

Johnson, S. (2020). 8 steps to content-first design. UX Collective. https://uxdesign.cc/8-steps-to-content-first-design-fa2885b9caee

Keshtcher, Y. (2021). How much do UX writers actually make? Salary survey report 2021. https://uxwritinghub.com/ux-writing-salary-survey-2021/

King, L. (2022). How content-first design creates a better user experience. Gather Content by Bynder. https://gathercontent.com/blog/designing-content-first-for-a-better-ux

Malik, O. (2003, Nov. 9). Gather them eyeballs. Dawn of the Micro Pubs. https://om.co/gigaom/gather-them-eyeballs/

Norman, D. & Nielsen, J. (2021). The definition of user experience (UX). Nielsen Norman Group. https://www.nngroup.com/articles/definition-user-experience/

Norman, D. & Tognazzini, B. (2015). How Apple is giving design a bad name. Fast Company. https://www.fastcompany.com/3053406/how-apple-is-giving-design-a-bad-name

R. Buckminster Fuller Quotes. (n.d.). BrainyQuote.com. https://www.brainyquote.com/quotes/r_buckminster_fuller_165610

Redish, J. (2010). Technical communication and usability: Intertwined strands and mutual influences. *IEEE Transactions on Professional Communication, 53*(3), 191–201. https://ieeexplore.ieee.org/document/5556477

Redish, G. & Barnum, C. (2011). Overlap, influence, intertwining: The interplay of UX and technical communication. *Journal of Usability Studies, 6*(3), 90–101. https://uxpajournal.org/overlap-influence-intertwining-the-interplay-of-ux-and-technical-communication/

Simon, H.A. (1971). *Designing organizations for an information-rich world.* Johns Hopkins University Press. https://digitalcollections.library.cmu.edu/awweb/awarchive?type=file&item=33748

Stafford, P. (2019, Aug 12). UX writer salary survey: How much money do UX writers make? *UX Writers Collective.* https://uxwriterscollective.com/how-much-do-ux-writers-make-salary-survey-revealed/

Verhulsdonck, G., & Shalamova, N. (2020). Creating content that influences people: Considering user experience and behavioral design in technical communication. *Journal of Technical Writing and Communication, 50*(4), 376–400.

Winters, S. (2019). What is content design? https://contentdesign.london/content-design/what-is-content-design/

2   The UX Writing Process

Chapter Overview

This chapter summarizes the iterative UX writing process as powered by the popular "design thinking" model. It begins by considering content as deliverables and how various requirements inform the design process. Then, the chapter offers an overview of the content lifecycle—from strategy to creation to evaluation—showing the multiple facets of content design and where UX writing is performed. The core of the chapter consists of an explanation of design thinking as a framework for content design. Traditionally, design thinking and content strategy are seen as mutually exclusive processes, but in this chapter, we show how they work together. The chapter provides a synopsis of the key phases of design thinking, which constitute the five chapters in Part 2 of the book.

Learning Objectives

- Recognize the distinction between content as a UX deliverable, a product, and a process.
- Describe the three continuums of UX writing as part of its taxonomy.
- Understand the lifecycle of content, from setting content requirements to evaluating content outcomes.
- Draw connections among multiple theories from adjacent fields to inform UX writing principles and practices.
- Understand the design thinking process and how it supports the UX writing workflow.

Content as a Product and a Process

We opened this book with an introductory chapter that spelled out the modern needs for user-centered content and how UX writing is born of those needs. User-centered content design is paramount to consumer satisfaction, brand loyalty, and, more importantly, equitable experiences that promote ethics and justice. As you may already know, there are many expectations for good content—timely, usable, useful, error-free, efficient, effective... you name it—and UX writers are tasked with creating content that is shareable, relatable, and pleasing. As well, the way we, as users, interact with devices and interfaces today calls for increasingly personalized content for different types of users and tasks or scenarios.

DOI: 10.4324/9781003274414-3

UX writing encompasses many different things wherein the content is crucial for facilitating a good user experience. Unlike the more conventional writing or technical documentation, UX writing is instead *writing in motion*, as in content development and implementation that are continuous and iterative in nature. UX writers work with shifting dimensions of content design. As content needs to be shareable, relatable, pleasant, and personalized to the individual user experience, writers cannot assume they are creating content only for a static page. Even though UX writers might be creating a "webpage," they can't assume the "page" will be an 8.5"×11" sheet of paper. Rather, they design for multiple screen sizes, devices, contexts, and user types. Depending on where users encounter the content, they may need different outputs based on the device they are using and their immediate needs. For example, a user may need directions to a new place they intend to travel to, listen to a digital voice assistant about the weather, or use an app while they are trying to do something. All of these require dynamic *microcontent* (see Chapter 9 for examples) that responds to the immediate situation of the user.

The attention to such omni-design aims to create an experience that resonates with the user and is pleasant and informative—what some marketing strategists call a *total experience*. And that experience can be with a physical product, a service experience (like ordering something from Starbucks), or an online interface, and it requires successfully managing the above processes in devising and delivering content.

Real World Snapshot 2.1: Content Design vs. Content Marketing

In the late 90s, Bill Gates famously coined the phrase "Content is king" to indicate how content is equally as important as the physical product it serves in our so-called attention economy (Papalamava & Garcia, 2019). People consume content as part of their everyday routine and culture. What's more, the social web affords and promotes the sharing and resharing of content, making content an invaluable yet immaterial commodity. This phenomenon turns many business strategists to the notion of content marketing—creating and distributing valuable, relevant, and consistent content to attract and retain a clearly defined audience—and, ultimately, to drive profitable customer action.

Nevertheless, is content marketing the same as content design? The short answer is "no." While content marketing aims to sell segments of a brand or product, content design serves the larger goal of creating and maintaining a desirable *user experience* for a product. Content design deals with strategy, as the content development agency Fractl (Milligan, 2019) puts it:

> Without a strategy, the content on your site won't be user-friendly or helpful to your readers. The strategy is needed to stay on-brand, to capture a voice, and to release content at the ideal time. Meanwhile, content marketing is what you need to elevate your strategy into generating leads and boosting your search rankings.

It goes without saying, both content marketing and content design need good content to achieve their respective goals. So, there are inevitable overlaps between marketing and design, including strategic planning and content promotion. Indeed, content may be king, but be sure to attend to other "royalty"—i.e., design and strategy—as well.

In the world of UX writing, content is the product, or, the deliverable. The UX of content directly affects users' perceptions of the brand, the organization, or even the entire industry. Many companies know the importance of paying attention to content design. As you will see in the variety of examples we cover in Part 2 of the book, these examples apply user-centered content design strategies to understand user requirements and emotional reactions to different scenarios involving the organization's product. One of such methods is "UX journey mapping"—a research-driven approach to visualize users' encounters, pain points, desires, and opportunities. You will learn more about this method in Chapter 5.

A user-centered approach to design can help companies develop seamless and continuously engaging products for users. Yet users grow along with content. What might be good content today may be undesirable tomorrow. For example, think of how Clippy in Microsoft Word has given way to users expecting to use voice commands on their phones to get relevant information that they need. All in all, since content is social, shareable, and relatable and needs to be personalized, engaging, consistent, and dynamic for users, UX writing is a challenging practice that is a continuously evolving field of practice.

But for the purpose of this book, we have to draw a more defined parameter for UX writing as a product as well as a process. As we describe in a later section here, UX writing combines solid principles of technical communication with human psychology, learning behaviors, human-technology interaction, rhetorical design strategies, and UX research into a total experience that is not only about the experience of the content interface but also that which is happening *around* the interface. It is also about knowing the exchanges between the user, device, and content and measuring this in a dynamic manner so it performs optimally in dynamic ways using common UX metrics for engagement, time-on-task, and task effectiveness.

As a process, UX writing depends on a recursive, iterative way of crafting and refining content in order to design something that is truly responsive to user needs. In this chapter, we will cover a cyclic model that UX writers use to develop desirable content. Before getting there, however, we are going to start by defining the scheme of UX writing.

The UX Writing Taxonomy: Three Continuums

Up until this point, we have not specified what UX writing looks like. If you've skimmed the table of contents for this book, you may notice that we dedicated Chapter 9 to identifying and expanding on the common applications of UX writing today. Here, we offer a rather generous overview of the UX writing taxonomy with help from current industry trends and U.S. federal government requirements. We synthesize these practices into three dimensions (Figure 2.1).

The first continuum we'd like to discuss is narrow vs. wide UX writing. While the narrow/wide UX continuum is surely more than binary (and not mutually exclusive), it provides a direct comparison between micro and macro components of UX writing. Narrow UX writing is concerned with accuracy and efficiency, both of which would immediately impact the experience of a user when using content. Narrow UX writing focuses on micro aspects of design that have to do with interactivity with an interface (e.g., buttons, menus, links), task assistance (like user guides, error messages), and information processing (web forms, logins). In Chapter 9, we provide an expansive list of examples to show you the common tasks involved in designing these content types. Narrow UX may also be specific to certain users or groups; thus, it could be considered localized or internal writing. An example of this would be style guides for a particular brand/product.

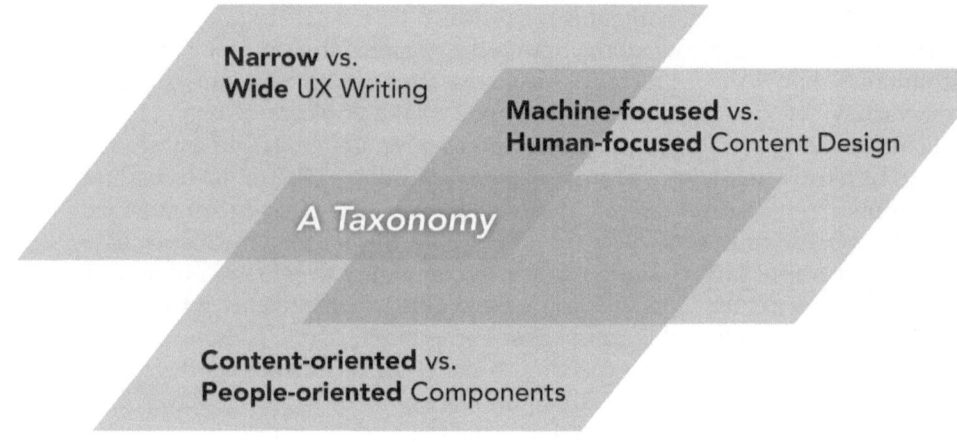

Figure 2.1 The three dimensions of the UX writing taxonomy.

Narrow UX is to trees what wide UX is to a forest. On the one hand, narrow UX writing deals with details. On the other hand, wide UX writing is concerned with global UX issues, such as international standards (like the ISO 9241-210),[1] regional or national culture, accessibility, and ethics. In other words, wide UX writing looks beyond situated aspects of design to cover big-picture matters that have strong generalization value across user populations and use contexts. For examples:

- Write content that does good and does no harm to anyone, intentionally or not.
- Avoid producing content that detracts from a content provider's value proposition.
- Ensure that every piece of content is accessible to everyone from anywhere.
- Create appropriate notices for users regarding privacy settings.
- Consider the digital carbon footprint. Foster sustainable consumerism to avert overuse.
- Be conscious of a given piece of content's effects on users' health.

These matters are external to an organization or product's convention. They guide writers and designers in creating content that serves a larger audience and has an ethical purpose.

The second continuum UX writers need to consider is machine-focused vs. human-focused content design. Similar to the first continuum, this non-exclusive comparison lies on a spectrum with some overlapping interests. Since the rise of artificial intelligence (AI) in computing, content has been consumed not just by human readers but also by "smart" agents that are programmed to detect language patterns and process linguistic data provided by people or other computing machines.

An example of machine-focused content design is writing for applicant tracking systems—software that helps companies organize job applications and select qualified candidates based on predetermined criteria. Yeqing Kong and her colleagues at North Carolina State University have studied the impact of AI-assisted automated recruiting technologies—including asynchronous video interviews, social media profiling, and neuroscience games—and noted that writers need to compose their materials (in this case, job application materials) for non-human readers that may weigh certain qualities (like credentials) greater than traditional human biases (like emotions).

For UX writers, this distinction is a kind of expanded audience awareness. It is not just about catering content for a particular readership but also recognizing how the "reading" method, or rather screening protocol, by AI agents may favor information differently than a human reader. The same awareness applies to conversation design in chatbots used across social media and many websites. Chatbot assistants can be incredibly cost-effective and time-saving for service-driven organizations, but UX writers must pay attention to the *balance* between talking to a human and talking with a human. AI chats need to sound natural, human, and better yet, humanized, which means that content should "get" human needs and emotional experiences. Figure 2.2 compares high and low "extraversion" in chatbot reactions to show the different effects they incite.

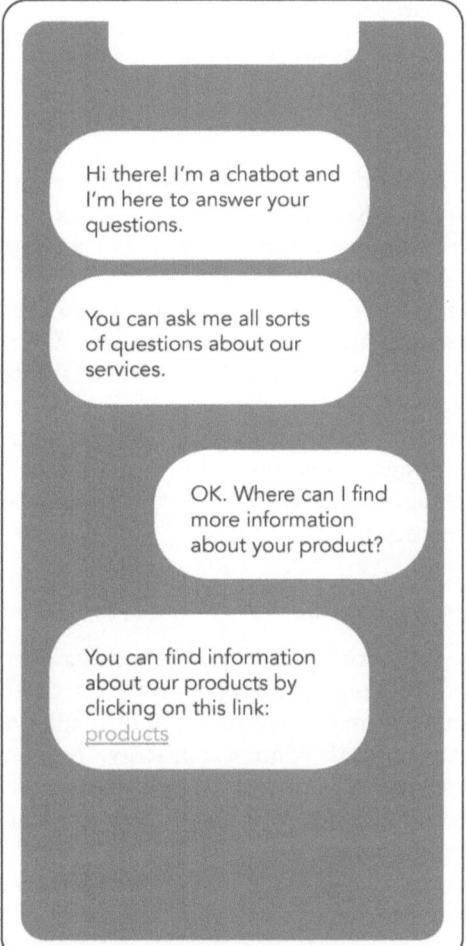

Chatbots with **high extraversion**:
- focus on social talk
- high use of exclamation marks
- use of emojis

Chatbots with **low extraversion**:
- focus on task at hand
- moderate use of exclamation marks
- no emojis

Figure 2.2 Create a personality for chatbots with desirable traits using high or low extraversion characteristics.

Source: Nerds & Company (2019).

Table 2.1 The "content strategy quad" devised by Brain Traffic, a content services company

Content-oriented components	People-oriented components
Identify Goals and Substance: focuses on what content is required to successfully execute your core strategy. It includes characteristics such as messaging architecture, intended audience(s), and voice and tone.	**Outline the Roles and Workflow:** focuses on how people manage and maintain content on a daily basis, including the roles, tasks, and tools required throughout the content lifecycle.
Determine Structure: focuses on how content is prioritized, organized, and accessed. Focuses on the content itself, including mapping messages to content, content bridging, and creating detailed page tables.	**Identify Policies and Standards:** focuses on the policies, standards, and guidelines that apply to content and its lifecycle, as well as how an organization will sustain and evolve its content strategy.

Sources: Usability.gov (https://www.usability.gov/what-and-why/content-strategy.html).

The third continuum is the comparison of content-oriented vs. people-oriented components in UX writing strategy. This existing comparison is made by the U.S. General Services Administration's technology transformation branch, which focuses on digital services. Relying heavily on Kristine Halvorson's (2011) work, this continuum separates the product from the producer (see Table 2.1). Content-oriented components include the goals of the content, which inform its structure. Whereas people-oriented components focus on workflow, roles, policies, and standards that inform the lifecycle of content design, which we will talk about next.

In sum, the taxonomy of UX writing provides three dimensions for understanding content design. Each dimension is a way into evaluating UX writing purposes, methods, and potential outcomes. UX writers should be able to discuss their approaches through these distinctions.

Content Lifecycle

Like any writing process, UX writing lives on a productive lifecycle that shapes the workflow of content creation, distribution, performance measurement, and improvement. By this point, you may have read about many content-related keywords or processes that sound quite identical: content "strategy," content "development," content "management," etc. These specific terms meant different parts of the design process. This section will piece together the puzzle using these terms and teach you the language for describing the UX writing process.

According to the U.S. General Services Administration (via usability.gov), there are five major stages in the lifecycle. In general, a content lifecycle includes the following:

- Audit and Analysis: Content stakeholder interviews, competitive analysis, objective analysis, and evaluation of the content environment (site, partner content, sister, and parent sites).
- Strategy: Determine topical ownership areas, taxonomy, process/workflow for content production, sourcing plan, voice, and brand definition.
- Plan: Staffing recommendations, content management system customization, metadata plan, communications plan, and migration plan.

- Create: Writing content, asset production, governance model, search engine optimization, and quality assurance.
- Maintain: Plan for periodic auditing, advise the client, and determine targets for success measures.

While these five general stages make up the backbone of the lifecycle, they contain more complexity when taking into account the need to evaluate and optimize content once it is implemented. To show a more expansive view of the lifecycle, we have mapped these stages onto the content circuit as shown in Figure 2.3.

The content lifecycle begins with research (i.e., auditing and analyzing). To create a set of requirements for content, which will serve as your overall project backlog, you and your team start by talking to users to gather insights and understand their needs. Then, you need to devise a strategy—a plan of action—to create the desired content. This strategy is informed by your motivations based on user research and the reality of your project (how much time and resources you have at your disposal). The actual crafting and designing of the content stage is known as content development. Often, industry practitioners use it interchangeably with content strategy. But we see planning and developing as two interdependent but distinctive stages. During the development stage, writers also create plans for managing the content as it is being developed. Content management typically involves

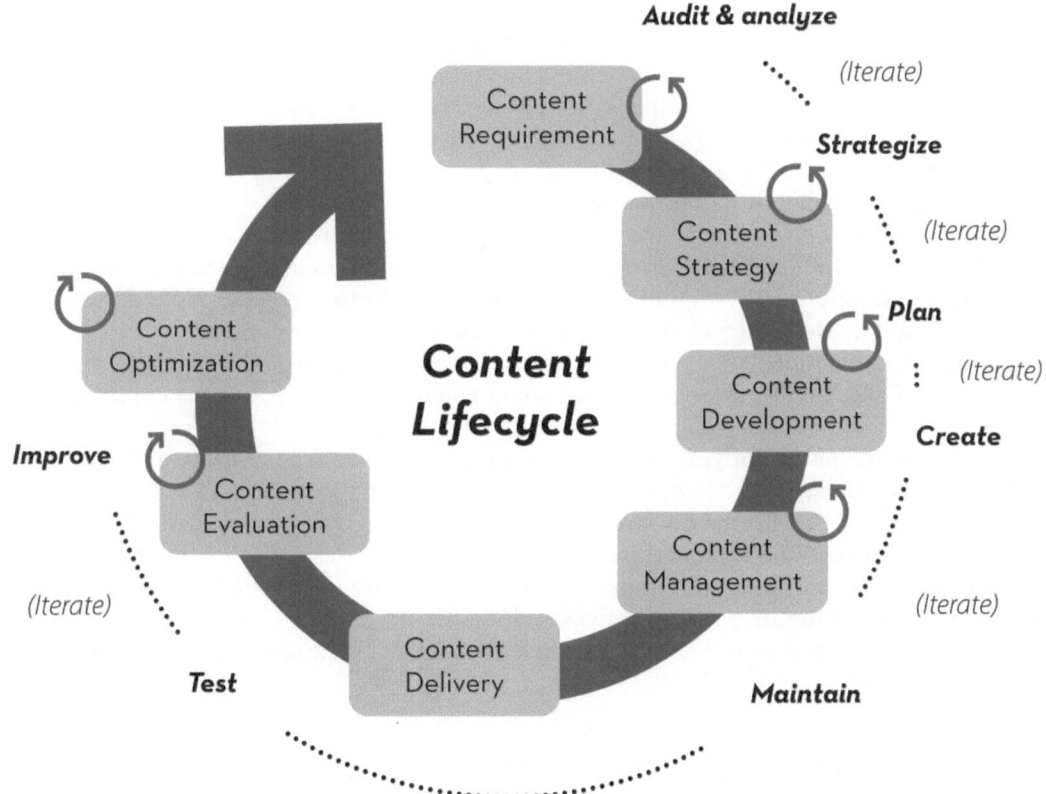

Figure 2.3 Content lifecycle, adapted and expanded from Usability.gov.

systems and platforms that support collaborative composition, single-sourcing workflows, and asset management. Note that this whole process is iterative, meaning that writers do not simply check off a list of to-dos in a linear approach but instead go back and forth on content research, strategy, planning, and creation.

When parts or all of the desirable content have been created and are ready to be deployed, you enter the next stage called content delivery. As the name suggests, delivery is about sending your content into the world so actual users can interact with it. This stage requires coordination between you (the design team), your organization's public-facing communication experts (i.e., public relations, spokespersons, etc.), and if relevant, the client who commissioned the content. Here's where we depart from the conventional five-stage model. Our content lifecycle does not end here. Immediately following the delivery stage is preparation for testing the implemented content. As indicated in Chapters 7 and 8 of this book, we advocate for a continuous improvement approach to content design. Once the content is out in the world, your team should track and measure the performance of the content using indicators that are appropriate for your context. With the findings about content use and users' reactions, your next focus is devising an optimization strategy to address arising concerns and leverage any emergent conditions that could be tweaked to benefit the performance of your content.

Here are some guiding questions for each stage of the content lifecycle:

1 **Content requirement:** What's our motivation for creating new content and/or updating existing content?
2 **Content strategy:** How should we go about creating and updating the content? What should be the project workflow (starting and ending points)?
3 **Content development:** What topic(s) should we focus on? What informs our topic decisions? How should we review content as it gets developed?
4 **Content management:** How should we manage the content being developed or edited? How might we collaborate on this activity?
5 **Content delivery:** How should we publish the approved content? How can we help users find the content? How might we localize the content for different audiences and contexts?
6 **Content evaluation:** How should we assess the impact of the content? How do we know when to update the content? How might we remove or archive content?
7 **Content optimization:** What are the concerns facing users brought about by our new content? What conditions are there that could enhance the success of the content?

Depending on the nature of your project, the content lifecycle may re-loop, and you might be taken back to a second, third, or fourth research phase to develop a new strategy and craft new content. As with the overarching mindset for design that we discussed in Chapter 1, the UX writing process is an iterative (cyclic, repetitive) process that favors improvement rather than merely checking milestones. Before we say more about the framework that we rely on to power an iterative process of UX writing, we need to look at several foundational principles built by experts and theorists from adjacent fields that have had strong influences on UX and writing practices.

Major Theories That Inform UX (and) Writing

UX as a practice has borrowed from social scientific disciplines and the humanities to cement the theoretical groundwork for describing and predicting human behaviors. As a modern profession, UX and writing benefit from other disciplines that also care about human interactions and how we make sense of things in the world. Here, we focus on four disciplines that we believe have the strongest influence on UX writing: psychology, human factors, rhetoric & philosophy, and business & project management.

UX writing understands human behaviors by learning from the field of psychology. Notably, American psychologist Don Norman (at the University of California, San Diego until 1993, and later the Nielsen Norman Group) has called attention to the use of cognitive science to study how people perceive problems and solutions in technological environments. Norman and Stephen Draper coined the term user-centered design in 1986, a project that gave birth to the UX profession. Norman's widely adopted book, *The Psychology of Everyday Things* (later retitled *The Design of Everyday Things*), applied the notion of conceptual and mental models to explain how humans learn to interact with interfaces and figure out unfamiliar situations. Other psychological theories—such as Gestalt, priming, and memory—are common principles used in UX design. Gestalt principles reveal how the human mind attempts to simplify and organize complex visual information (a popular example is how the mind's eye will fill in the missing parts of an image to create a whole, which we will see in the discussion of eye-tracking in Chapter 4). Priming effects are used in design to mentally prepare the user for subsequent effects or tasks. For example, if you tell a kid about red cars before going on the road, they are likely to notice more red cars during the trip.

Memory is arguably the most important consideration for UX designers and writers because it directly affects the presentation of information. Cognitive load is the amount of mental effort required to form a person's working memory (what's needed to do an active task). Since the human brain prefers to conserve energy, the less effort needed to perform a task, the better. This is why UX writers present information in chunks or categories, which, according to cognitive psychology, helps reduce cognitive load for information processing and improve short-term retention of the information. As human memory degrades due to aging, UX writers also need to consider other strategies to create content that is accessible to older users.

Besides concerns about the brain and the mind, UX professionals need to pay attention to human physical—or rather, physiological—factors. This is why in many psychology departments, there is an area of emphasis called "human factors" and "ergonomics," an offshoot of UX that focuses specifically on the human body and psychomotor issues related to voluntary and involuntary actions. These actions can affect a user's productivity, safety, and error prevention. For UX writing, human factors principles can guide the design of content usability, such as button display, font sizes, colors, and paragraph length for interfaces of different sizes.

Within the North American context, UX largely resides in language and writing departments, creating a close affinity between UX and the rhetorical tradition. Rhetoric, as one of the classical areas of formal education, is concerned with discourse and persuasion. It is a study of techniques for informing or motivating people in a given situation. For more than 2,000 years since its invention, this ancient art of communication is still prominent today in UX writing because rhetoric teaches practitioners to identify the available means for influencing people for specific purposes. Rhetoric also appeals to philosophical concerns such as ethics and virtue, which are increasingly important to UX writing given the cultural and political climate our society is experiencing. UX practitioners apply rhetorical principles to design effective, engaging, and ethical content.

While we are among the many advocates who use UX design to achieve social good, the UX industry still primarily serves commercial needs. Writers and designers learn from modern business and project management methodologies to facilitate feasible—and more importantly, profitable—workflow. Project development models such as Agile, Scrum, and Lean management (see Chapter 3 for more) are popular models used in UX writing workflows to foster collaboration and cross-functional interactions. Alas, even the content lifecycle that we presented earlier is affected by the ideologies in these project management models. Another eminent practice in UX writing that is informed by business management is structured authoring, or single-sourcing, which aims to save time, money, and resources. Structured authoring applies a modular content creation and management approach to design (learn more in Snapshot 2.2). Business schools, "bootcamps," and professional development centers are significant sites of influence on UX writing today.

Real World Snapshot 2.2: Structured Authoring for UX Writing

Jojo is a writer at SimsTech, a virtual reality (VR) world-building service that specializes in virtual architecture and interior design. Jojo's role at the company includes documenting the specs and developer guide for ST-4XR (SimsTech's most recent VR sim operation system) and creating content to attract potential clients.

During Quarter 1 of 2021, Jojo produced a series of documentation for the ST-4XR that has received praise for its clarity and usability. Jojo is quite happy about the work she's done.

In Quarter 4 of 2021, Facebook announced the release of Metaverse, a network of 3D, immersive social spaces. SimsTech has since received an abundance of client requests to build "worlds" in the metaverse rather than isolated VR. To do so, SimsTech updated ST-4XR to ST-4XR-Meta, which included additional plugins to integrate with Meta.

Jojo was asked to update all previous ST-4XR documentation to show the Meta integration. With more than 450 entries in the previous documentation and specs and about 200 of those needing updates, Jojo found herself dreading the need to go over all the entries and then edit the necessary ones. What's more, SimsTech has already planned another system update in Quarter 3 of 2022, which may then require Jojo to do another documentation overhaul. As Jojo sat down to brainstorm a more efficient way of doing this seemingly impossible mission, she took the week off to attend the annual summit hosted by the Society for Technical Communication that happened to be held in her city.

At the summit, Jojo was introduced to the practice of structured authoring. Structured authoring is a standardized approach to writing documentation where the content is controlled by pre-defined rules. Using XML (eXtensible Markup Language), the writer marks up the content according to what it semantically represents. Figure 2.4 shows how structured content can be tagged using semantic tags that reflect what kind of content it represents.

The structure and semantic tagging enable and simplify content reuse, increasing both consistency and efficiency in technical documentation. Structured authoring also enables and simplifies the automation of publishing technical content and applying formatting separately.

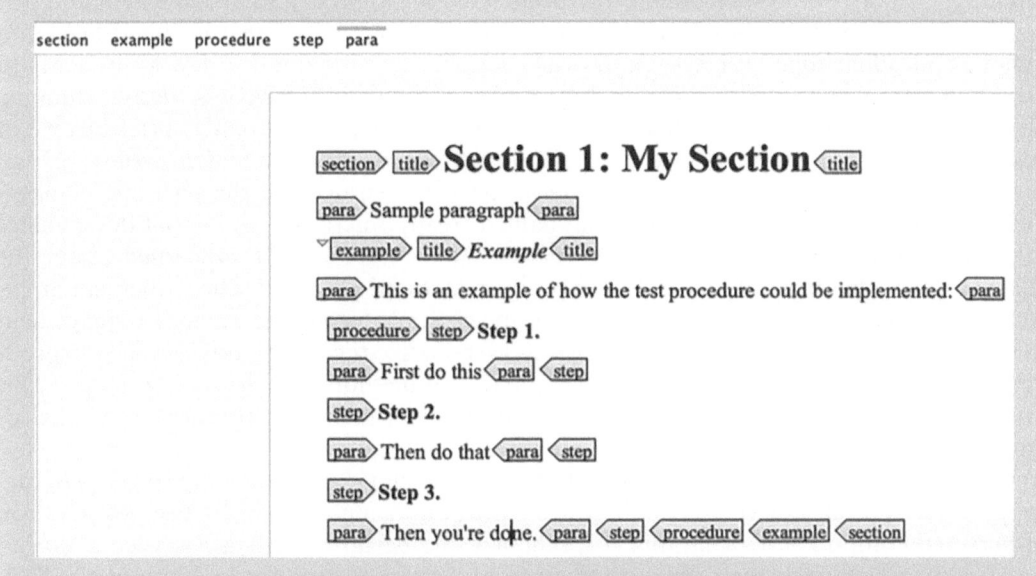

Figure 2.4 XML tags are used to categorize content to make it easier to locate and edit after it's been published.

Source: Paligo (2020).

To put it simply, you have 100 manuals in the company, and all of them contain the same text fragment. If one of the manuals is changed, you would only need to update that one file, and whenever you opened and published the 100 documents, they would all automatically pull the changes.

Jojo was over the moon to have learned that structured authoring may save her time and labor in updating her VR documentation now and preparing for future updates as well. She has since been a structured authoring evangelist to her fellow writers and colleagues in the field.

In sum, UX writing takes advice from many adjacent fields, as we have briefly covered them here. It goes without saying that as UX grows as a profession, it will cultivate its own principles and best practices. However, at this time, UX writing relies on theories derived from other disciplines to inform effective design. You may notice that we have reserved the arguably most influential discipline of *design studies* for the next section. Design studies deserve its own dedicated space, as it's inseparable from UX writing. Next up, we discuss the relationship between design and UX writing as a segue into the central framework that upholds the premise of this book—i.e., design thinking.

From Design Science to Design Thinking

Earlier, we previewed the *iterative-ness* of UX writing through the content lifecycle. We have mentioned that this iterative process is powered by a design framework that supports user-centered content solutions. This design framework, popularly known as "design thinking," was an evolution from the Taylorian scientific approach to management and was heavily influenced by the Scandinavian participatory design movement in the 1980s. At a time when leading-edge businesses and organizations were looking for new approaches to inventing products as well as new angles to market their innovativeness, design thinking offered an affordance to doing both. In this section, we capture a (really) brief history of design thinking to give you its essence and rationale for its adoption by UX professionals.

Taking heed of the industrial revolution in the 19th century and the scientific management theory of Frederick Taylor that had gained steady traction in industrial design since the beginning of the 20th century, modern designers were largely concerned about the efficiency and usefulness of any designed product. Around the same time, American architect Buckminster Fuller coined the term "design science" to combine Taylor's management theory with human factors and psychology to systematize the design process. The byproduct of design science is a structured, efficacy-centered methodology for doing design. This historical ideology has continued to shape the thinking of many practitioners when considering the purpose of design in society.

Fast forward to the global recession of the 1980s when mass marketing of utopian ideals persevered alongside the technological revolution brought about by the proliferation of personal computers. During this era, scholars and design theorists observed a "social turn" in practice given the growing influence of critical theory across humanistic disciplines, including design studies. Designers were concerned about the political, socioeconomic, and systemic power of design beyond its functional purposes. It was also during this time that designers began to realize the social impact of their design work. Whether it's building a transit station, charting a world map, crafting a sports team logo, writing a newspaper headline, or inventing a wireless (mobile) phone, designers paid closer attention to how their choices affect social norms and expectations. As well, they were enticed by the rapidly changing landscape of information technology, in part due to the rise of tech giants like Apple, IBM, Microsoft, and others that were steering the trends of design in the 1990s. In 1991, when the three-way merger took place between David Kelly (professor at Stanford University), London-based designer Bill Moggridge, and Mike Nuttall of Matrix Product Design, the resulting design firm, IDEO, became an evangelist of design thinking—a departure from the scientific method into socially rooted changemaking powered by a set of designerly values. These values were further promoted by the Stanford educational offshoot of IDEO, namely the Hasso Plattner Institute of Design (fondly known as the d.school).

Design thinking was dubbed a socially responsive way to innovate solutions thanks to the popularization of Nordic participatory design methods. Challenging the expert-driven (scientific) approach to innovation, design thinking leverages individual experiences and collective expectations to create more user-centered solutions. As a way of structuring the design process, design thinking specifies five distinctive phases:

1 **Empathy:** Understand user behaviors and needs in the context of their work and lives.
2 **Definition:** Unpack and translate user research findings into compelling insights.
3 **Ideation:** Generate radical design solutions and alternatives.
4 **Prototyping:** Get ideas into the real world through material or digital means.
5 **Testing:** Get feedback on the designed solution.

Figure 2.5 illustrates the five phases in a cyclic manner. This illustration emphasizes the iterative nature of design, where linearity is overcome by recursive activities. In what follows,

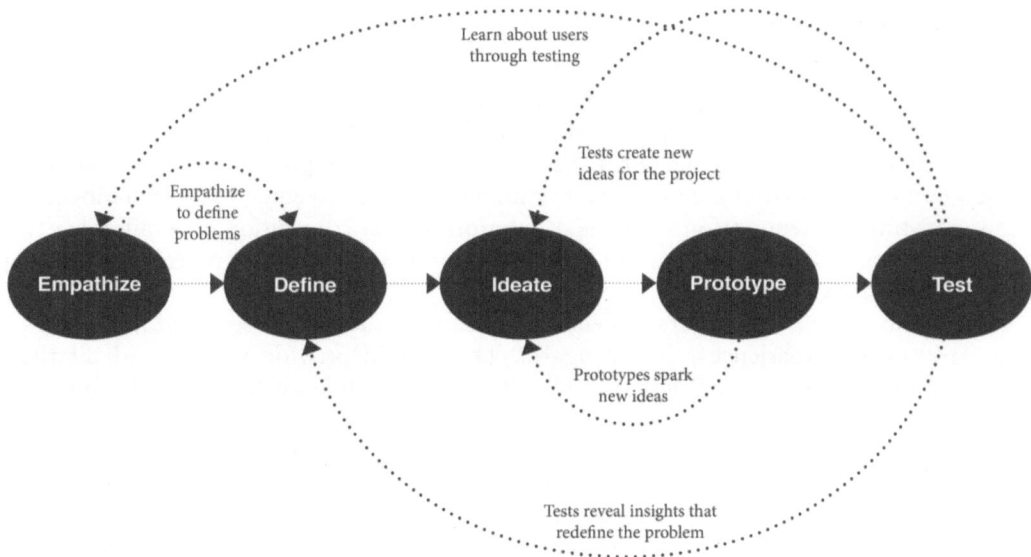

Figure 2.5 The five basic phases of the cyclic design thinking methodology, adapted from the Stanford d.school's model.

we situate these phases in the UX writing process as a way to demonstrate how design thinking supports user-centered design.

More than just a method for innovating content, design thinking also promotes a set of attributes that the Stanford d.school characterizes as a "mindset" for design:

- **Show, don't tell:** Communicate your vision in an impactful and meaningful way by creating experiences, using illustrative visuals, and telling good stories.
- **Focus on human values:** Empathizing with the people you are designing for and taking feedback from these users are fundamental to good design.
- **Craft clarity:** Produce a coherent vision out of messy problems. Frame it in a way to inspire others and fuel ideation.
- **Embrace experimentation:** Prototyping is not simply a way to validate your idea; it is an integral part of your innovation process. We build to think and learn.
- **Be mindful of process:** Know where you are in the design process, what methods to use in that stage, and what your goals are.
- **Bias toward action:** Design thinking is a misnomer; it is more about doing than thinking. Bias toward doing and making over thinking and meeting.
- **Radical collaboration:** Bring together innovators with varied backgrounds and viewpoints. Enable breakthrough insights and solutions to emerge from the diversity.

(D.school, 2011, p. 3)

We believe these attributes of design thinking can benefit UX writing because the mindset complements the phases of the design process while reminding writers about making their ideas explicit (visual and experiential), centering empathy as a core value, tinkering with radical ideas and collaborative processes, and enabling action-driven problem-solving (rather than abstract).

Phase 1: Empathizing with Users and Understanding Their Needs

In the words of former first lady Michelle Obama, empathy is an action. It is not just thinking about someone but interacting with them to understand their situations. To empathize is to put aside one's assumptions about another person and to truly learn from their experiences. In UX writing, when writers are presented with opportunities to update or create new content, the first phase of design thinking urges them to enter a research mode. This means resisting the temptation to name a solution due to convention or availability, and instead entering into a listening and observation mode in order to explore how everyday people experience content.

During this phase, UX writers are encouraged to interact with users and inquire about their daily tasks, experiences, and struggles. This would provide writers with firsthand accounts of users' desires and needs. In Chapter 4, you will learn the methods commonly used in such user research to collect insights, including contextual inquiry, goal-setting interviews, user story mapping, and surveys. It is important for writers to set aside their personal biases toward certain solutions during this phase because the goal of empathizing is to collect real-world, empirical information that can guide the design direction.

If you ever feel compelled to prioritize certain ways of solving a problem, consider the exercise of asking yourself how your own positionality, power, and privilege have affected your decisions. Researchers Natasha Jones, Kristen Moore, and Rebecca Walton (2016) have provided a useful heuristic for recognizing one's own complex circumstances and the effects they have on our responses to different situations.

- **Positionality:** How different aspects of identity are markers of relational, historical, and dynamic qualities that inform our attitudes and behaviors.
- **Power:** One's ability to influence others through dominance, exploitation, or subjectivity.
- **Privilege:** An advantage that benefits those who are granted this status to the exclusion and detriment of those who are not.

For UX writers, this first phase of design thinking is an opportunity to scrutinize structural challenges—cultural, economical, ethnic, racial, social, and others—and the way they affect content solutions in the world today. When in doubt, always ask an actual user how they feel rather than guesstimating their needs or their vocabulary.

Phase 2: Defining Problems and Opportunities

The next phase of design thinking asks writers to bring together the knowledge and information they have gathered from their empathizing activities to create a proper definition of the problem at hand. At this stage, writers and their teams collaborate to interpret user responses to their surveys, observations from situational observations, and any key insights generated from user research. The goal here is to craft a focus to guide writers in their content design process.

The methods used in this phase include persona building and journey mapping, which are demonstrated in detail in Chapter 5. Writers create images and descriptions of the user using research data to form tangible references that can help them keep the user in mind while designing. These personas are used in drawing the user journey maps in order to visualize the realistic experiences users undergo when engaging with content. As part of this definition phase, writers seek to specify the parameters of their project to enhance clarity for the whole team, thus enhancing the quality of decision-making during design.

Phase 3: Ideating Content Solutions

The core of design thinking is innovation. The human-centered design philosophy underscoring design thinking calls for radical innovations that are creative yet grounded in social realities. During the ideation phase, writers practice different methods of "flaring" to generate unconventional, unlimited ideas. Strategies such as card sorting and affinity diagramming are common in design workshops to promote an understanding of users' mental models. Mental models are how people make sense of the world around them, including their interactions with information and tools. These activities can show writers how people think, learn, and apply prior knowledge to solve problems.

A signature characteristic of design thinking is its participatory approach to ideation. Also called co-designing, methods such as the participatory design workshop and user-led exercises are encouraged in the ideation process to create solutions that are most desirable to users themselves. In Chapter 6, we give you an overview and multiple examples of such methods so you can practice integrating them into your ideation process.

Phase 4: Prototyping Content

Design thinking favors a bias toward action. Integral to a content design process is the active creation of physical or tangible models for the ideas generated from the ideation phase. This next phase, called prototyping, emphasizes the importance of bringing ideas into the material world. If you're new to UX design, you may be intimidated by the term, but it is not as technical or complex as it sounds. A prototype is a preliminary (proto) model (type) of an idea. Significant to user-centered design, prototyping is a revealing process; it serves to expose aspects of a design that may be hidden in hindsight. Prototyping motivates writers to look at their content solution from multiple angles once it's been brought into existence through proper structures and arrangements.

While most content solutions manifest in digital forms, design thinking encourages you to build low- or high-fidelity prototypes to allow for user interaction with your ideas. Low-fidelity prototypes, such as paper prototypes and sketches, let you and your users try the ideated solution early—and often—during the design process. Higher-fidelity prototypes that are typically built for formal usability testing are more polished and improved versions of your idea in tangible form. In Chapter 6, we offer tips and tools for effective prototyping, including mockups and wireframing.

Phase 5: Testing and Validating Solutions

Prototypes are instrumental for testing the ideated solutions and validating the usefulness of your designed content. During the testing phase of design thinking, writers recruit real users who are representative of the project's target audience to participate in various testing activities. The focus of these tests is on observing users' reactions to the design and how they make sense of it. In Chapter 7, you will learn the ways to set up synchronous/asynchronous, moderated/unmoderated tests that would yield significant results about your design. You may collect qualitative, quantitative, attitudinal, and behavioral data from these tests to understand both the usability and the UX of your content.

The testing phase in design thinking tends to reveal the messiness—or iterative-ness—in design. It is common for writers to return to definition or ideation after learning about user reactions from testing. This helps them further refine their design or address something that was overlooked previously.

Tracking and Measuring Content Performance

The design process doesn't end with the post-testing phase. When a version of the content is ready to be implemented—or shipped, in the practitioners' lingo—it is tracked with specific analytics tools. This lets writers learn about its performance in the wild. There are various tracking methods and tools, some of which we review in Chapter 8. Using these tools and the key-performance indicators set by writers and clients, content performance is captured and analyzed to determine future iterations. Thanks to sophisticated data analytics packages such as Google Analytics that use different metrics, you can automate much of the tracking and measurement process with relative ease.

Managing the Iterative Workflow and Continuous Improvement

All throughout this chapter, we have stressed how UX writing should be an iterative and cyclic practice. Per design thinking's mindset and methodology, content design should exercise experimentation, evaluation, and revision. In the content lifecycle, we have included not just content evaluation after deployment but also content optimization. Together with

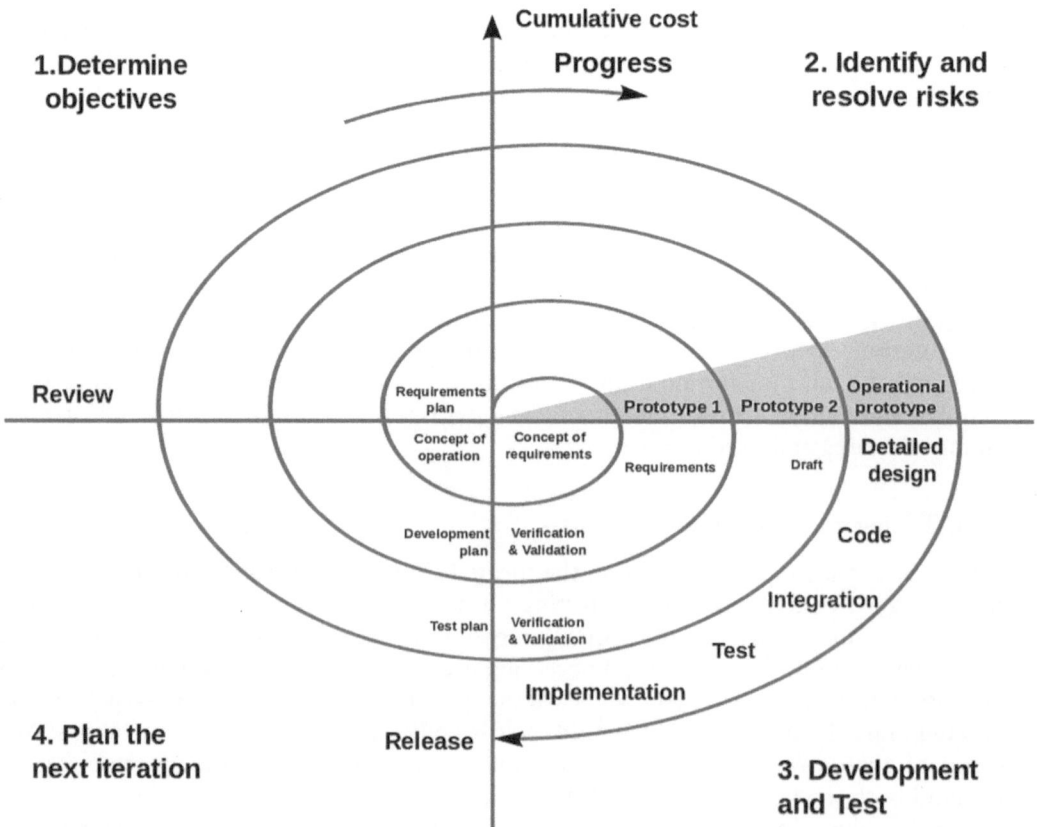

Figure 2.6 The spiral model for iterative development.

Source: Wikimedia (inspired by Boehm, 1988).

the iterative workflow of design thinking, this kind of continuous improvement approach requires your team to devise a plan for managing iterations and optimizations.

There are no set guidelines for devising such a plan. However, we recommend a time-tested model for considering the iterative design and continuous improvement process (Figure 2.6):

- **Plan:** What are the content requirements?
- **Determine objectives:** What should the content achieve, ideally?
- **Evaluate alternatives:** What are the risks and opportunities for new content?
- **Prototype:** What might the content look like?
- **Validate:** How might the content be developed, deployed, integrated, and accepted by users?

With a plan, you and your team can set priorities at different stages of your development cycle and be guided by the objectives of each stage. Just remember, UX writing is always contextual, so your plan should be responsive to the unique situation and challenges that make up the backdrop of the project. A continuous improvement plan is subject to your scope of work and the viability of your goals.

Conclusion

UX writing is dynamic. It is a product and a process that involve:

- Identifying content requirements and user needs.
- Designing and managing content that can be used and reused across different and multiple channels, devices, and modalities.
- Developing content in teams through a continuous improvement workflow using the design thinking framework.
- Strategizing content for UX across contexts so the overall/total experience is seamless and pleasant.
- Tracking content to see how it performs over time.

In this chapter, we have discussed the taxonomy of UX writing via three key continuums and unpacked the content lifecycle in UX writing. You have been introduced to the major theories informing UX and writing practices, as well as the design thinking framework that underscores our thinking in this book.

Chapter Checklist

UX writing can be powered by design thinking to address the ever-evolving, shifting technological environments that affect the user experience. UX writing involves content as a dynamic product that requires strategic development and management through multiple activities.

- *Treat UX writing as dynamic—rather than static—to create shareable, relatable, and pleasant content.*
  - View UX writing as a product.
  - View UX writing as a process.

- *Understand the taxonomy of UX writing.*

  - Narrow UX is concerned with accuracy and efficiency; wide UX focuses on global issues.
  - Machine-focused content design is writing for non-human readers; human-focused content design is writing that creates personable experiences.
  - Content-oriented components are goals and substance that determine content structure; people-oriented components are roles and workflow that influence content policies and standards.

- *Follow the content lifecycle to manage the content design process.*

  - Audit and analyze content requirements.
  - Devise a content strategy based on insights.
  - Follow the strategic plan for content development.
  - Use appropriate tools for content management.
  - Coordinate your efforts for content delivery.
  - Test the implemented solutions through content evaluation.
  - Track and enhance your solution via content optimization.

- *UX writing borrows theories and principles from adjacent disciplines.*

  - Cognitive psychology teaches writers about human decision-making and meaning-making.
  - Consider memory and cognitive load when presenting information.
  - Human factors considerations can guide the design of content usability to reduce error, enhance performance, and ensure safety.
  - Rhetoric teaches practitioners to identify the available means for influencing people for specific purposes.
  - Project development models are used in UX writing workflows to foster collaboration and cross-functional interactions.

- *Design thinking underscores the mindset and methodology for UX writing.*

  - Empathize: Understand user behaviors and needs in the context of their work and life.
  - Define: Unpack and translate user research findings into compelling insights.
  - Ideate: Generate radical design solutions and alternatives.
  - Prototype: Get ideas into the real world through material or digital means.
  - Test: Get feedback on the designed solution.
  - Use design thinking attributes to complement UX writing ideology.

- *UX writing is iterative.*

  - Exercise experimentation, evaluation, and revision to enable continuous improvement in content design.

Discussion Questions

1 In what ways is UX writing a product as well as a process at the same time?
2 Describe, in your own words, how the design thinking process and values support the goals of UX writing.
3 What key deliverables occur at each stage of the content lifecycle?
4 Discuss what "iterative" means to content design and why it is important.

Learning Activity

One way to learn about the UX writing process is by reverse-engineering an existing content design. Take, for instance, how a new Netflix movie release is designed—what gets considered when a new title is ready for the public and how the release is ideated. Discuss with a friend:

1 What are the potential content requirements for the new Netflix movie? Consider content other than the movie itself—the release announcement, social media engagement, public posters, celebrity interviews, award nominations, book tours, talk show appearances... What else?
2 How might the requirements turn into an actionable content strategy?
3 How might the strategy fit into a content lifecycle that takes care of the full, iterative process of content design?
4 In what ways might the new Netflix release be tracked and measured for content performance?

Write down a list of challenges you have with the above steps and use them as guiding questions when reading the following chapters.

Notes

1 ISO 9241-210:2019—Ergonomics of human-system interaction — Part 210: Human-centered design for interactive systems (https://www.iso.org/standard/77520.html).

References

Boehm, B.W. (1988). A spiral model of software development and enhancement. *Computer, 21*(5), 61–72. http://www-scf.usc.edu/~csci201/lectures/Lecture11/boehm1988.pdf

D.school (2011). Design thinking bootcamp bootleg. https://dschool.stanford.edu/resources/the-bootcamp-bootleg

Halvorson, K. (2011). Content strategy and UX: A modern love story. *UX Magazine.* https://uxmag.com/articles/content-strategy-and-ux-a-modern-love-story

Jones, N., Moore, K., & Walton, R. (2016). Disrupting the past to disrupt the future: An antenarrative of technical communication. *Technical Communication Quarterly, 25*(4), 211–229.

Kong, Y., Xie, C., Wang, J. Jones, H., & Ding, H. (2021). AI-assisted recruiting technologies: Tools, challenges, and opportunities. In E. Lane and H.M. Lawrence (Eds.), *SIGDOC '21: The 39th ACM international conference on design of communication* (pp. 359–361). ACM. https://dl.acm.org/doi/proceedings/10.1145/3472714

Milligan, A. (2019). Content strategy vs. content marketing: What's the difference? Fractl. https://blog.frac.tl/content-strategy-vs-content-marketing/

Nerds & Company, (2019). How to personalize chatbots: 3-step personalization model. Chatbots Life. https://chatbotslife.com/how-to-personalize-chatbots-3-step-personalization-model-3385c803580

Norman, D. (1988/2013). *The design of everyday things* (revised and expanded edition). Basic Books.

Norman, D. & Draper, S. (1986). *User centered system design: New perspectives on human-computer interaction.* Lawrence Erlbaum Associates. Paligo (2020). About structured authoring. *Paligo Documentation.* https://paligo.net/docs/en/about-structured-authoring.html

Papalamava, I. & Garcia, K. (2019). Content strategy 3.0, a new winning formula. *BCG Digital Ventures.* https://medium.com/bcg-digital-ventures/content-strategy-3-0-a-new-winning-formula-5aa6a1edf5de

3 Building a UX Writer Outlook

Chapter Overview

In this chapter, we highlight eight essential traits you need for a successful and lasting career as a UX writer. These traits, based on human-centered, socially sensitive design philosophies, are the values, mindset, and professional practices you need to design user-centered content. By following them, you keep your focus on the user and the user's experience to generate content that is relevant, useful, and usable.

Learning Objectives

- Understand the influence of design thinking characteristics on a UX writer's outlook.
- Recognize the significance of human-centered design principles in UX writing.
- Characterize the practice of UX writing in terms of social awareness.
- Describe the desired qualities needed to succeed in UX writing careers.

Traits: Your Tools of the Trade

In the previous chapter, you were introduced to the design thinking mindset and methodology. This chapter will demonstrate how the mindset translates into likable traits for UX writing. Let's begin with a mini-design exercise (Figure 3.1).

Scenario: A traveler is in an airport waiting for the last leg of a flight home when their flight gets abruptly canceled due to bad weather.
Challenge: Write a message from the airline app notifying the traveler of the cancellation and what they need to do next.
Consideration: This is a sensitive situation with a high chance of aggravating the user. Consider using an empathy-driven message that provides assistance to the user so they get peace of mind.
Solution: What would you devise? In the space below, write a short notification message for the traveler and discuss your rationale with a friend.

The keyword in this mini-design exercise is *empathy*. In the last chapter, you learned about the user-centered content development workflow that is called design thinking. What makes

DOI: 10.4324/9781003274414-4

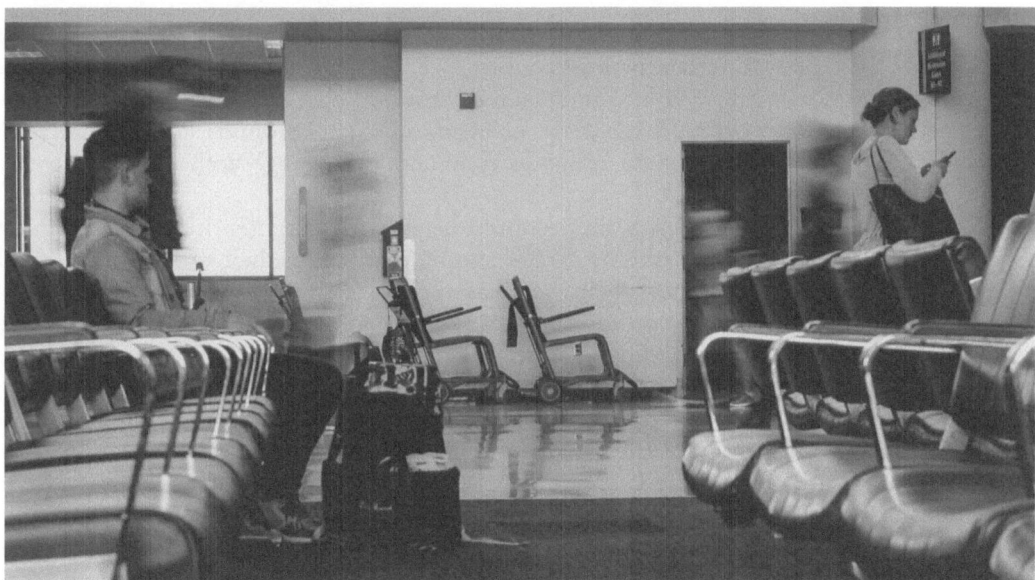

Figure 3.1 Passengers waiting at an airport gate.

Source: Andrik Langfield on Unsplash.

design thinking so popular and widely deployed in product development and content strategy is its insistence on basing design decisions on user needs and requirements. To truly practice writing *for* user experience, UX writers need to empathize with their audience so they can facilitate a content design that is pleasing and helpful to the user.

So, back to the mini-design scenario above, the message from the airline app needs to acknowledge the existing emotional experience of the user given the cancellation of their flight. Instead of simply giving directions to the user to book a later flight, UX writers will first communicate apologetic messages and make compensation to the traveler prior to instructing them how to get on the next flight. This may appease the traveler, reduce complaints launched against the airline, and help avoid bad reviews that could harm future business.

Empathy is only the beginning. In this chapter, we will take a look at a total of eight traits that make for a successful UX writing professional based on observations made in the current state of our digital society and advice gathered from industry experts who have worked with content design teams. These traits are:

- Listening empathetically
- Having a strong cultural awareness
- Being digitally savvy
- Knowing when to break the rules
- Advocating for ethical practices
- Practicing Agile collaboration
- Seeing the forest as well as the tree
- Mentoring others

Listening Empathetically

As Michelle Obama put it, empathy is an action. To empathize with someone, you need to engage in activities that allow you to understand your audience's contextual conditions—where do they live and work? What's their socioeconomic background? How comfortable are they with various technologies? How do they actually engage with your content or product? How do they feel about it?

To answer these questions, UX writers should listen with the intention to get to know their audience. This means that writers do not simply read about their users from the web or assume certain characteristics of their users based on biases or prior conceptions. Unfortunately, this happens more often than the average person expects in the workplace (see Snapshot 3.1). For reasons pertaining to time, resources, and availability, some practitioners opt not to engage directly with the users they are designing content for. Rather, good UX writers participate in listening activities that expose them to the real user experience. You will see in Chapters 4 through 8 the various methods and processes that can help writers gather user information, needs, and requirements effectively. Here, we provide some tips for establishing this first important trait as a UX writer.

UX professional Maximillian Schmidt (2020) said there are five levels of listening: Ignoring, pretending, selective, attentive, and empathetic listening (Figure 3.2). The progressive levels show how someone could be listening but not caring. To truly empathize with users, as a human-centered design principle, writers and designers need to step out of their "frame of reference," which means actively reminding themselves how their worldview, background, and personal values can filter the things they see or hear. To achieve empathetic listening, UX writers need to listen without judgment, check their understanding against individual frames, and allow their audience to express how they feel in their own ways.

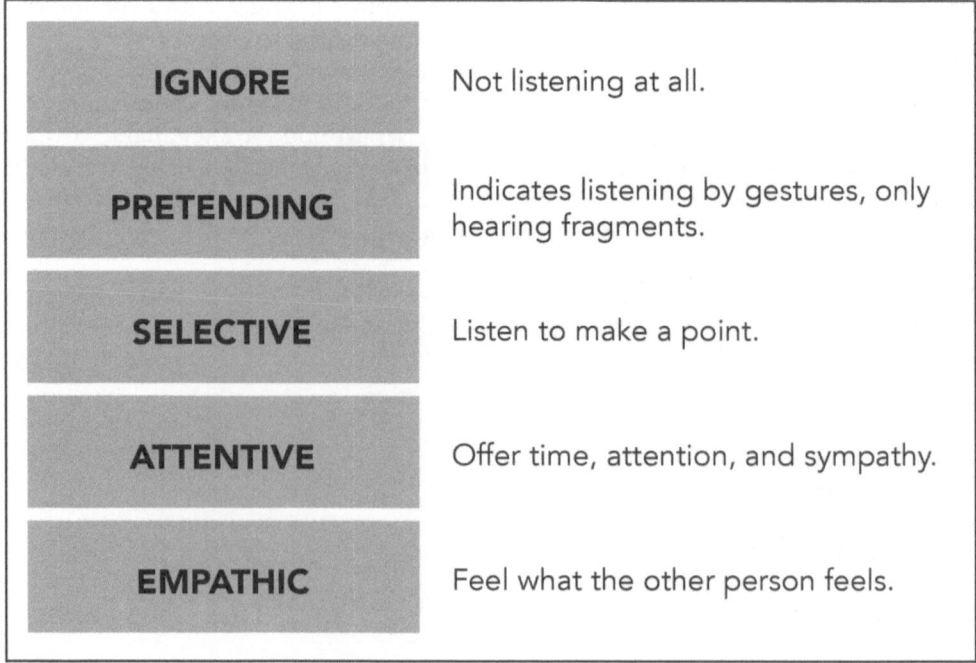

Figure 3.2 Schmidt's five levels of listening.

Now why is this sort of listening so important to the UX writing process? As noted earlier, many practitioners do not have the time or space to slow down and listen. The cost of ignoring the user's feelings, however, is certainly greater than the investment of time. Empathetic listening will give you the most authentic experiences from the users. The better you know your users, the more likely it is that you will be able to create products that serve their needs. In the long run, this creates favorable results and acceptance by users, which will generate brand loyalty—and that's good for business!

Real World Snapshot 3.1: Maximilian Schmidt's Account of Developers' Pushback

No matter what I as a designer propose, developers always just say, it is impossible and costs a fortune. Of course you then show astonishment to that user test you set up. Have you instead ever tried to understand what the problem really is, then asked the developer what they would suggest to achieve the same results? I was surprised so many times that the solution might be even easier. Learn from other's expertise and experience.

Schmidt's experience here is typical of those who work in product development and think user research is a luxury. No doubt, many small agencies and organizations consider it a luxury rather than an actual requirement due to limited resources. But as Schmidt's account shows us, listening is a critical part of the process that shouldn't be skipped simply because of convenience or wanting to save money (Figure 3.3).

Figure 3.3 Talking with users is the first step to empathetic listening.
Source: LinkedIn Sales Solutions on Unsplash.

To listen is to learn. And learning takes you far. Whether you aim to become a UX researcher or content designer, listening can only benefit your work and your understanding of the user. Such understanding will prove to be necessary for creating products that people actually want to use.

Having a Strong Cultural Awareness

Good listening helps writers cultivate awareness of differences. Having sensitivity toward diverse cultures and social perceptions is particularly important for UX writing in this global market. As the internetworked world brings people closer in shared content spaces—like user forums, social media, and online community events—writers need to display the crucial trait of recognizing how users from around the globe understand and react to certain content.

Culture is a sticky subject, and it takes a lifetime to master, even for seasoned professionals. There is geographic culture, ethnic culture, religion culture, discipline or field culture, workplace culture, and even family culture. Culture can be as global as how we read stop signs on the street and as local as how we organize our food pantry or clothes closet at home. Building an awareness toward the different kinds of cultures for the purpose of UX writing takes practice and exposure. Thus, it is helpful for aspiring writers to immerse themselves in new or unfamiliar environments to experience and learn how people work in those spaces.

You may ask: Well, how might I do that... Specifically, to build a strong awareness of cultures? The practice of developing cultural awareness is a lifelong endeavor, to be sure. However, for UX writers, the key practice is spending time with people, places, and activities that give you the opportunity to gain insights into how different settings affect user behaviors, experiences, and outcomes. For example, you may travel if you can or engage in professional events like international conferences and conventions to meet and hear from people outside your immediate circles. As the world was forced to "go virtual" during the recent health pandemic, you could also leverage the opportunities to join events and activities that are held across the globe from your local office to participate in conversations that could open your own world to the diversity and multiplicity of cultures.

The bottom line is to break outside of your existing experience—your frames of reference—so you can be more conscious of the way of life that may be different from your own. UX writing needs this kind of awareness because writers need to relate to audiences and users who aren't always the most immediately relatable to them. One's interpretation of colors, for example, is very much informed by their cultural upbringing and ongoing experience. Thus, successful UX writing can only be brought about by a keen insistence on learning others' cultures and creating products that speak to the understanding and expectations of these cultures.

Cultural awareness expands beyond ethnicity, race, and a person's biological background. Culture also includes certain ways of life that people subscribe to, such as music taste, preferred entertainment genres, languages, and technology use (and non-use). For instance, Sam is a 25-year-old middle school teacher who enjoys using Twitter to share her everyday musings with her followers. Based on this description, a good writer will be able to identify several key cultural factors that may inform Sam's experience with the social media application. In designing content that caters to someone like Sam, writers apply their understanding of different cultural elements to ensure a desirable user experience.

Being Digitally Savvy

The third trait of a good UX writer is digital savviness. UX writing requires constant engagement with modern technologies. While writing technologies revolve mainly around the typical word-processing software or content management application, writers who wish to be ready for the next big thing in UX and content design need to equip themselves with the capacity to learn new tools quickly. This is a trait that most commonly shows up as a desired qualification for UX writing jobs. Employers seek candidates who display the promise to pick up unfamiliar tools and tinker with new, more effective ways of writing.

What does it really mean to be savvy? While there are no standard metrics to measure how savvy someone is, many industry practitioners apply the Gartner Hype Cycle for emerging technologies to locate where different kinds of technology users enter the scene of new adoption as a way to indicate their comfort level with unfamiliar tools. Essentially, the cycle observes five key phases of a technology's lifecycle, from inception to adoption in the mass market (Table 3.1).

Digitally savvy users typically engage with new technology during the "trigger" phase, when expectations for the new technology are on the rise. Early adopters, sometimes called lighthouse users, are crucial in the technology emergence process as they are usually the ones who provide candid critiques on new features or activities. Early adopters consider themselves savvy at learning unfamiliar designs, which makes for a great trait in UX writers. Of course, you don't always need to be the first to catch a new technology, but knowing the emerging design trends can help your team plan for future content strategies before they become commonplace (Figure 3.4).

Table 3.1 The descriptions of key phases in the Hype Cycle (Chaffey & Ellis-Chadwick, 2016).

Phase	Description
1 Technology trigger	A potential technology breakthrough kicks things off. Early proof-of-concept stories and media interest trigger significant publicity. Often, no usable products exist, and commercial viability is unproven.
2 Peak of inflated expectations	Early publicity produces a number of success stories—often accompanied by scores of failures. Some companies take action; most don't.
3 Trough of disillusionment	Interest wanes as experiments and implementations fail to deliver. Producers of the technology shake it out or fail. Investment continues only if the surviving providers improve their products to the satisfaction of early adopters.
4 Slope of enlightenment	More instances of how technology can benefit the enterprise start to crystallize and become more widely understood. Second- and third-generation products appear from technology providers. More enterprises fund pilots; conservative companies remain cautious.
5 Plateau of productivity	Mainstream adoption starts to take off. Criteria for assessing provider viability are more clearly defined. The technology's broad market applicability and relevance are clearly paying off. If the technology has more than a niche market, then it will continue to grow.

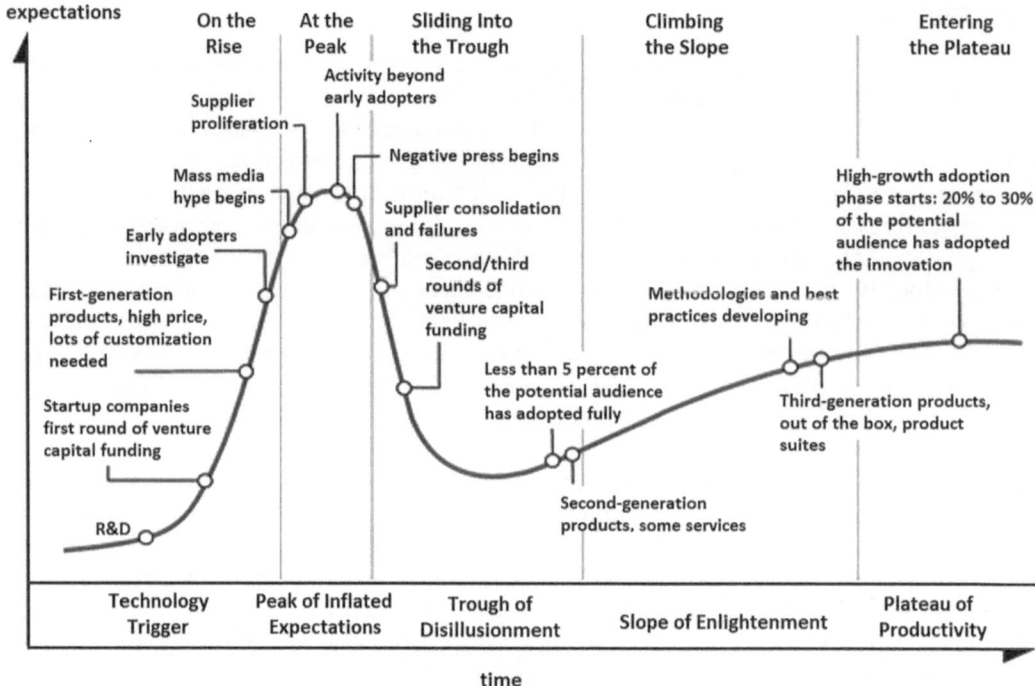

Figure 3.4 The Gartner Hype Cycle for technology adoption.

Source: Olga Tarkovskiy on Wikimedia Commons. CC BY-SA 3.0 https://creativecommons.org/licenses/by-sa/3.0/ deed.en

The one constant about technology is that it never stays the same. Because technologies always evolve and improve, good UX writers are invested in continuous learning so they can continually pick up a new tool, a process, and a methodology to support their professional growth. It may be cliché to say that a lifelong learner is a successful learner, but a UX writing career depends on this mantra to nurture practitioners who are eager to improve their skill sets and expand their toolkits. Fortunately, there are many venues for novice and experienced writers alike to immerse themselves in continuous learning. Online course providers like Coursera, Codecademy, edX, and LinkedIn Learning are excellent resources for you to learn cutting-edge tools and new trends in digital applications to keep up with current practice. Global companies like Google and Meta are also offering insider courses that provide timely training on proprietary tools (like Google Analytics and Meta Business Manager). These spaces also let you meet other aspiring practitioners who seek to develop their technological proficiency and thus an opportunity for you to create a community of inquiry.

Ultimately, the platforms, tools, and standards—really, any technology—will change to accommodate our ongoing lives; the key is to stay open to learning. A digitally savvy UX writer is one who is willing to try out new apps and content design methods so they can compare and contrast what's new with what's now, and select the best way of managing their workflow.

Knowing When to Break the Rules

Evolving technologies call for constant attention to change. Similarly, UX writers need to focus on current trends in communication and user behaviors so they can facilitate appropriate content design. While this is a textbook for UX writing, and a textbook traditionally teaches readers the "rules" or best practices for engaging a subject matter, we now break the textbook rule to posit that a good UX writer should also know when not to follow the conventional practice. Knowing when to break rules may sound like a cliché at work, but it is indeed an important distinction that separates good UX writing from great ones.

The most common rule often broken by UX writers is grammar. Consider the popular song by The Beatles:

She's got a ticket to ride.
She's got a ticket to ride.
She's got a ticket to ride.
But she don't care.

It's a wonderful tune and rhythm, but the lyrics don't conform to traditional grammar rules. If they did, the song would not have rolled off the tongue that easily. It would have cost The Beatles billions of dollars in sales, too. In all seriousness, this is an important UX move that song writers, singers, and music producers have deliberated and committed to for the purpose of prioritizing experience over rules.

UX writers will always find themselves in instances where they must make decisions to craft clarity over dogmatic systems for writing (e.g., grammar, punctuation, spelling, citation format, etc.). We are not saying that these systems are bad. They do provide scaffolding for clear communication, which is important for transactions of information. Yet, in certain scenarios, these scaffolds can come in between the intentions of communication and the actual experience of the audience. In fact, the U.S. federal government has even made a law to enforce clarity in any legal and governmental communications that are public-facing. This law, the Plain Language Act, was signed by President Obama on October 13, 2010.

Real World Snapshot 3.2: First Rule of Plain Language

Rule #1: Think about Your Audience
One of the most popular plain language myths is that you have to "dumb down" your content so that everyone everywhere can read it. That's not true. The first rule of plain language is: *write for your audience.* Use language your audience knows and feels comfortable with. Take your audience's current level of knowledge into account. Don't write for an 8th-grade class if your audience is composed of PhD candidates, small business owners, working parents, or immigrants. Only write for 8th graders if your audience is, in fact, an 8th-grade class.

Make sure you know who your audience is—don't guess or assume.

Plainlanguage.gov (2011). *Federal plain language guidelines.* p.1. https://www.plainlanguage.gov/media/FederalPLGuidelines.pdf

In UX writing, clear writing means designing content that best serves the needs of the user with direct language and effective use of space. When in doubt, err on the side of being conversational in your writing style. This will make your content more relatable and comprehensible. Wherever applicable, humor is typically welcomed to paint a more human personality for a brand. (Caveat: don't overdo humor. It can be awkward.) In the Application Exercises near the end of this chapter, you will find a writing challenge where you can practice "breaking rules" to increase the clarity of the given message.

Advocating for Ethical Practices

Using plain language can help make your content more comprehensible to the non-expert user, as our previous section reveals. It is, in fact, an ethical design practice. To write for user experience is to advocate for people in ethical ways. UX writing reflects our social thought, values, and culture. It is important for writers to practice ethical methods when creating content. The first ethical consideration is accessibility. Writers are responsible for making sure everyone gets the same access to the content they create. This means building alternative pathways for users to get to the content rather than only favoring the popular way. For example, UX writers use high-contrast interfaces and alt-texts to allow low-vision or blind users to access the content with screen readers (an assistive technology that renders texts and images on a page into speech or braille output). Captions and subtitles are made for similar accessibility concerns.

Second, UX writers need to be sensitive toward inclusivity in content design. Make sure your work does not discriminate against anybody, whether by intention or not. Don't repeat this Facebook incident that has put the company in a poor light:

> A video of two Facebook employees using a soap dispenser in a company washroom went viral on Twitter. The footage revealed that when a white employee used the soap dispenser, he received the soap easily. But when a black employee put his hand under the sensor-driven soap dispenser, he got no soap at all—even after multiple attempts! The reason: the sensor wasn't able to sense his hand! This is a typical example of how people's racist biases can impact human-machine interactions if product developers and designers fail to recognize such possibilities.
>
> (Ghanchi, 2021)

UX writers serve as advocates for ethical design. They remind designers that the product must be usable by all. Another ethical concern in UX practices is dark pattern design. Dark patterns are features made to deliberately trick users into buying things or giving away information they don't intend to. Unfortunately, dark patterns are becoming pervasive in UX design. You have probably seen them on subscription pages, online shopping websites, and other digital ads that lure users into undesired content via clickbait. There is a list of common dark patterns on the website darkpatterns.org (courtesy of Kiess, 2019). UX writers should develop a keen eye toward these disguising practices and help designers recognize them as unethical methods.

Practicing Agile Collaboration

One effective way to mitigate some of the ethical problems presented earlier is for UX writers to work in Agile collaboration with designers. "Agile" here refers to an iterative, continuous, and cyclical approach to project management. Besides the common UX research and writing skills required for content design, Agile collaboration and project management are the topmost sought-after proficiencies in UX writing positions. As UX writers participate in almost every phase of the product development process, they need the ability to partake in team decision-making and give feedback to multiple stakeholders. This is why a lot of design organizations deploy structured collaborative methodologies to ensure a common language for writers and designers to talk about products.

The most popular methodology is Scrum. Inspired by rugby, this Agile collaboration process has gained prominence in software development since the late 1990s. It creates specific roles for writers and designers to manage a project, such as the product owner, scrum team, and scrum master (who is responsible for organizing various scrum events). The scrum events serve to create explicit requirements for a project (the backlog), the timeline for project development, the "sprint" meetings, and the review or evaluation criteria for the design. Scrum, of course, is one way to facilitate collaboration. Many other methods, such as Lean, Six Sigma, and even Scaled Scrum, are available options for design teams. The idea is to have a set methodology and vocabulary for teams to manage their interactions.

Another key feature of Agile methodologies is visible thinking and planning. The Kanban method, which borrows from the Japanese term for "vision board," is a signature part of many Agile collaboration structures where teams maintain a visual planner for their project. A standard Kanban consists of three panels—(1) to do, (2) doing, and (3) done. This simple yet useful way of visualizing project status helps keep teams informed about current and future activities (Figure 3.5).

Having an understanding of how an Agile methodology works would place you in favorable light as a UX writer. This trait is becoming increasingly desirable by hiring managers because the workplace, UX and beyond, is evolving to include cross-functional and multidisciplinary collaboration. An Agile approach gives project teams a proven strategy for managing research as well as creative activities in a collaborative environment. If you can demonstrate knowledge of this modern approach, you'd certainly stand out as a candidate for UX writing jobs.

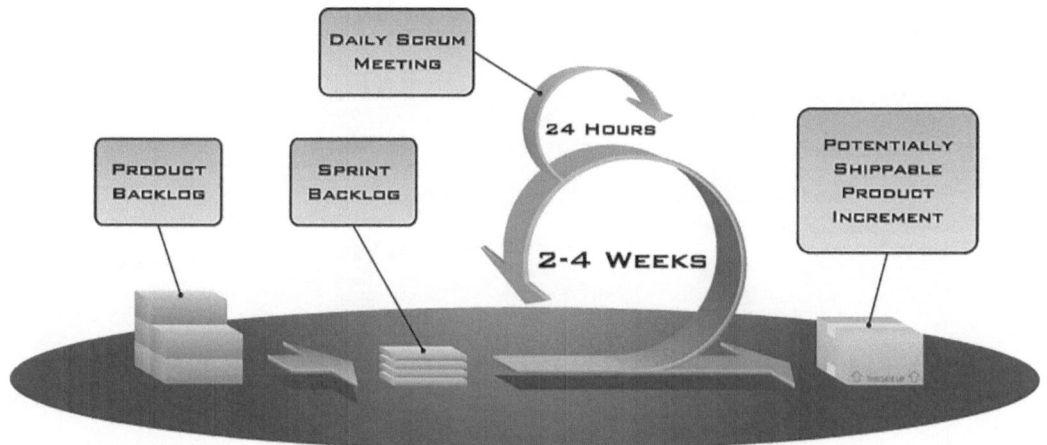

Figure 3.5 The Scrum methodology for Agile collaboration.

Source: Mountain Goat Software on Wikimedia. CC BY 2.5 https://creativecommons.org/licenses/by/2.5/deed.en

Seeing the Trees as Well as the Forest

UX writing professionals are big-picture thinkers. This core trait means that writers pay close attention to details in their work while being able to see the great impact of their work on a project or even the industry they serve. Good writers create UX content for multiple audiences, stakeholders, and constituents. They understand business strategy, customer experience, company mission, and industry trends. Here is an account by Nir Yuz, UX Studio Manager at the growing web design provider Wix:

> Our job as UX experts is a balancing act. We're always trying to figure out the right combination for all our ingredients: UI, UX, graphics, marketing and other business goals. I sometimes get frustrated when, in order to achieve optimal UX results on complex products, I need to sacrifice some of the cooler design elements.
>
> (Pedzai, 2016)

Thankfully, as we have alluded to in the Agile collaboration trait, UX is inherently an iterative process that requires writers to engage in a nonlinear workflow. This cyclical workflow can ensure that teams pay attention to multiple scenarios, potential outcomes, and likely and unlikely reactions by different users. Iterative testing and prototyping in this process can allow for show-stopping mistakes to be identified early and often (Figure 3.6).

To practice seeing the forest amidst the trees, so to speak, UX writers can apply literal checklists for big-picture items when working on projects so they can remember the larger implications of their work. Ask questions such as:

- How will this design affect targeted as well as unintended users?
- How will this content be maintained or updated over the next few months and years?

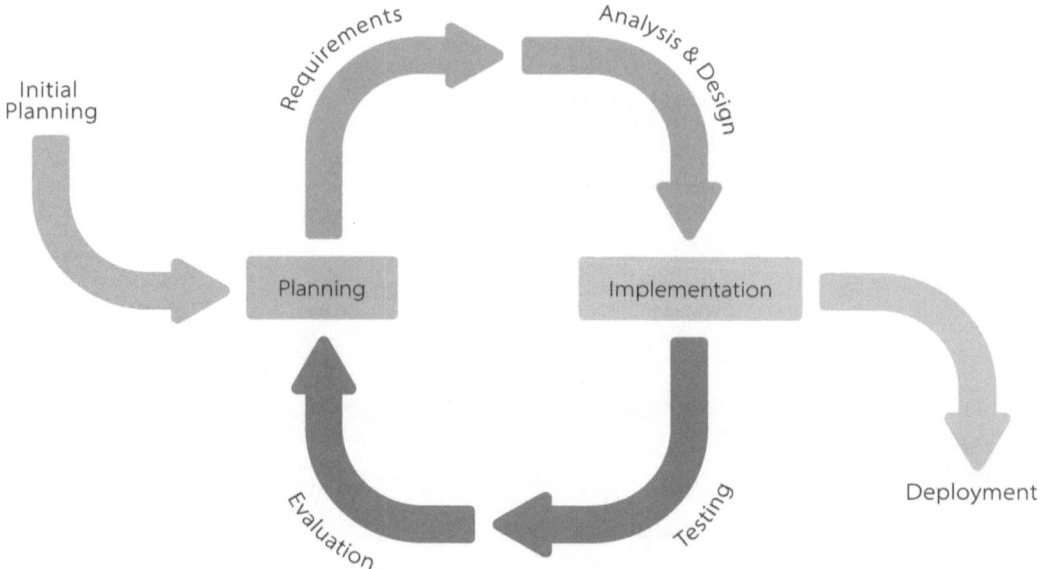

Figure 3.6 An iterative process for UX product development.

Source: Krupadeluxe on Wikimedia. CC BY-SA 4.0 https://creativecommons.org/licenses/by-sa/4.0/deed.en

- What safeguards will be in place to protect the company's reputation?
- How will this design influence industry trends?
- In the unlikely case of a show-stopping event, how will we react?

UX impacts business. UX influences how people perceive a brand, an organization, and a technology. When writing content for UX, you and your team will need to determine concepts and solutions that best serve the larger needs of the product. Just like Nir Yuz puts it, UX is a balance between getting optimal results and achieving business goals. Being a big-picture thinker will serve writers well in this complex role.

Mentoring Others

Last but certainly not least, a key trait of a great UX writer is sharing their knowledge to help other writers succeed. For a small but growing profession, this trait is imperative to the future of UX writing. A lot of UX writing activities happen within the silos of design companies and the confines of particular projects, thus making their presence less prominent than those the public may be more aware of, like health sciences, journalism, engineering, and business management. Yet, UX writing permeates all parts of our lives, and without it, there could be catastrophic mishaps in our daily activities. So, as practitioners, we need to foster a better recognition of the importance of UX writing in the modern workplace.

Thankfully, UX writers are already a generous community of practitioners who value co-mentorship and learning from each other. Even as we were developing this book, the three of us benefited from the resources provided by UX writers who shared their experience, advice, case studies, and feedback on design practices. A number of UX writing communities we recommend for those who are new to the profession are:

- UX Collective - https://uxdesign.cc
- UX Planet - https://uxplanet.org
- Career Foundry - https://careerfoundry.com/en/mentors
- UX Professionals Association - https://uxpa.org
- Association for Computing Machinery Special Interest Group in Design of Communication - https://sigdoc.acm.org
- Nielsen Norman Group - https://www.nngroup.com/articles
- Interaction Design Foundation - https://www.interaction-design.org/community

These communities welcome curious minds as well as seasoned practitioners alike to discuss the work of UX writing, the prospects of the career, and how to manage challenges. No doubt every project is different and requires unique propositions; having a community of support is helpful to decision-making processes, especially when others have success stories that you may want to emulate.

Now, mentoring is a two-way street. As you become more familiar and proficient with UX writing, the field counts on you to extend your expertise to those who are new to it. Sharing is caring. To continue the culture of mentoring, our community needs people to contribute their expertise and experience to those who seek it. One way to share is by blogging ideas via open-access platforms or personal websites. Another popular way to mentor others is to participate in local or regional meet-up events or conventions to formally present your work and teach others how to succeed in UX writing. No matter what level of mentorship you choose to engage in, the key here is to maintain service to the community so the field can continue to grow in meaningful ways.

Conclusion

Now that you've been given the tips on the traits that make for an ideal UX writing professional, what are the top three traits you feel most prepared for? In the space below, write out your proudest accomplishments.

What traits would you like to further develop? How might you do so? Write down some aspirational goals and tactics for reaching those goals.

Next up, we will begin a new part of the book where you'll be introduced to the myriad of methods and approaches to empathize with users, define content requirements, ideate user-centered solutions, prototype content that addresses specific needs, test the effectiveness of your design, and track and measure the success of your content using analytical tools. Just like the first trait we opened this chapter with, we are going to start with methods for empathy.

Chapter Checklist

To succeed in UX writing, a few "traits of the trade" can be helpful as you navigate this emergent career opportunity. In this chapter, we have zeroed in on eight distinctive attributes of good UX writers that can serve as guideposts for those who aspire to excel in this work.

- *Listen empathetically to understand the users' needs.*
  - Recognize your personal frame of reference.
  - Let your users tell you about their true experiences.

- *Cultivate a strong cultural awareness to create a desirable user experience.*
  - Immerse yourself in new environments and learn how people live and work.
  - Attend meetings or conventions that let you meet new people.

- *Be digitally savvy with new and emerging technologies.*
  - Catch emerging technological trends, perhaps by being an early adopter.
  - Learn new tools via continuing learning platforms.

- *Know when to break rules in writing or design.*
  - Craft clarity over prescriptive grammar.
  - Use plain language to increase comprehension.

- *Always advocate for ethics on behalf of users.*
  - Ensure accessibility and inclusivity.
  - Advocate for ethical design.

- *Work with teams through Agile collaboration.*
  - Learn Agile methodologies.
  - Practice working in cross-functional teams.

- *Keep the big picture as well as the details in mind.*
  - Create checklists for macro-deliberations.
  - Look beyond the parameters of the project.

- *Mentor others.*
  - Participate in communities of practice.
  - Share your own knowledge and experience.

Discussion Questions

1 Upon reviewing all of the UX writer traits highlighted in this chapter, share with a friend or colleague which of these qualities you already possess and how you can display them in your professional outlook.
2 Which of these traits do you feel are lacking in your current outlook, and how do you plan to develop them?
3 In what ways do these traits supplement or replace other common "soft skills" or qualitative skills in the workplace?

Learning Activity

The Amazon Mechanical Turk (MTurk) is a crowdsourced surveying platform that attracts the general public to sign up for human intelligence tasks (HITs) such as testing products or answering interview questions. Anyone can create tasks on MTurk and pay both the participants ("Workers") as incentives and MTurk for the hosting service. Below is the pricing information page on MTurk. Readers may find it challenging to understand how much MTurk would actually charge them based on the kind of workers they require.

Pricing

Pay Only for What You Use

The price you (the Requester) pay for a HIT is comprised of two components: the amount you pay Workers, plus a fee you pay Amazon MTurk, which is based on the amount you pay Workers. Additional details are as follows:

Worker Reward

You decide how much to pay Workers for each assignment.

MTurk Fee

20% fee on the reward and bonus amount (if any) you pay Workers.
HITs with 10 or more assignments will be charged an additional 20% fee on the reward you pay Workers.
The minimum fee is $0.01 per assignment or bonus payment.

Masters Qualification

There is an additional fee for using the Masters Qualification
The fee is 5% of the reward you pay Workers.

Premium Qualifications

There is an additional fee for using Premium Qualifications
The additional fee per assignment starts at $0.05 and varies depending on the qualification.

Rewrite the page with more direct copy and examples to increase clarity. Provide a short reflective memo to justify your writing strategy and choices made.

References

Chaffey, D. & Ellis-Chadwick, F. (2016). *Digital marketing: Strategy, implementation, and practice* (Sixth ed.). Pearson.

Ghanchi, J. (2021). Ethical design: Why is it critical for UX designers? UX Matters. https://www.uxmatters.com/mt/archives/2021/02/ethical-design-why-is-it-critical-for-ux-designers.php

Kiess, C. (2019). The good, the bad and the ugly of ethics in UX design. Prototypr. https://blog.prototypr.io/design-ethics-c739bcf511fe

Pedzai, C. (2016). Is UX design a good fit for you? 8 personality traits you must have. UX Planet. https://uxplanet.org/is-ux-design-a-good-fit-for-you-8-personality-traits-you-must-have-7b3cee71c6c4

Plainlanguage.gov (2011). Federal plain language guidelines. P.1. https://www.plainlanguage.gov/media/FederalPLGuidelines.pdf

Schmidt, M. (2020). The many levels of listening: Empathic listening as a design superpower. UX Collective. https://uxdesign.cc/empathic-listening-32ec5ad955cb

Part 2

Processes

This section zeroes in on the five core phases of design thinking, starting with **Chapter 4,** where we emphasize the importance of empathy and how it manifests in the practice of user research. After learning about ways to learn about people's attitudes and behaviors with content, you can find methods for translating user needs into actionable insights for content design in **Chapter 5.** In **Chapter 6,** you will learn and exercise ideating content solutions and making them into tangible, testable forms via prototyping. **Chapter 7** teaches you how to test prototyped content and manage its deployment to the public. Then, **Chapter 8** covers various technical and analytical tools to track content performance once it is live.

DOI: 10.4324/9781003274414-5

4 Empathizing with and Assessing User Needs

Chapter Overview

In this chapter, we focus on the "empathizing" phase of design thinking. We present six methods to help you listen to, observe, empathize with, and learn from the users for whom you are designing content as a UX writer. We begin by helping you have a productive kick-off meeting with the client during which you set goals, responsibilities, and schedules. Understanding the six methods we cover here will help you and your client plan a successful project.

In addition to goal-setting with clients, the other five methods you will learn here are:

- Contextual Inquiry, where you watch, listen, and talk with users as they do their work in their own work settings.
- User stories, which can come from contextual inquiry or other sources and which give you users' identities, needs (goals), and reasons for the need.
- Task-based usability testing (aka "think-aloud protocols), where you watch and listen as users work with a product (great as a technique to understand changes that are needed in an existing product—as well as very important throughout design and development).
- Eye-tracking that you can include in task-based usability testing.
- Strengths, weaknesses, opportunities, and threats (SWOT) analysis, where you and others involved in a project come up with and discuss an organization's SWOT to success.

Chapter 4 situates UX writing in design thinking philosophy and user-centered design processes, particularly through the "empathizing" phase of design thinking.

Learning Objectives
- Understand how research with users develops a UX writer's empathy for users' needs and goals.
- Use goal-setting interviews to establish and prioritize users' goals.
- Conduct contextual inquiries to observe how users' contextual environments impact the use of a product's design.
- Create user stories and story maps to help design teams develop a shared vision of a "minimum viable product" from an end user's perspective.

DOI: 10.4324/9781003274414-6

- Set up and conduct "think-aloud protocol analyses" to observe users setting goals and completing tasks with products in real time.
- Understand the basic principles of how eye-tracking systems function, the types of data they can report, and important limitations on how to interpret eye-tracking data.
- Conduct SWOT analyses to develop a snapshot of an organization's past, present, and future content strategies.

Starting with Empathy

Empathy is an action-based principle of design thinking that can be carried out with UX methods. In UX writing, empathy guides content design by requiring writers to practice user research before and during the design process. In Snapshot 4.1, you can learn about a UX writer's role and some key research practices in her position. Throughout this chapter, we present foundational industry methods for understanding users and user needs as the starting point for UX writing.

Real World Snapshot 4.1: Vignettes of UX Writers' Work Environments: Maggie's Job

Maggie works for a large, financial services organization that primarily provides processing services for credit card companies, banks, and lending institutions. Her company has sites all over the country, and because their customers require that a wide variety of accounting information be collected and processed in large databases, Maggie's company has several software and application development teams constantly working both to create new software packages for different types of emerging financial services as well as updating old software to work securely in cloud storage environments, to support new operating systems, or to update network security protocols, just to name a few types of upgrades.

At any given time, Maggie's company can have 15–20 major software and application projects underway, and each of these projects will be supported by different types of cross-functional teams. As a result, Maggie's company is divided into departments specializing in the areas of expertise needed. For example, some teams may need several software engineers with expertise in PHP coding and MySQL database management, someone with subject matter expertise in the accounting processes used, someone with graphic design experience who can create visuals for the graphical user interface that the team is developing, a project manager to coordinate the activities of the whole team, and so on.

Each member of a team reports directly to the supervisor of their department and is likely to be serving on several different project teams at the same time. Maggie, for example, works in the UX Research department; her job title is "UX Designer." She has three other colleagues in her department with whom she can share ideas, and she is currently working on four different projects with four different project managers. Even though everyone on the project teams works for the same company, Maggie "bills her time" to the projects on which she is working, so that a portion of the project manager's budget for the project goes to Maggie's department to compensate her department for Maggie's salary, travel, and indirect overhead costs (like her desk, computer, phone, etc.). As a result, before a project begins, Maggie and her director meet with the project manager and create a memo of understanding ("MOU") that details the scope, amount of time, and type of work Maggie will be performing on the team. Maggie's activities can vary a lot from project to project.

Even though Maggie's title is UX Designer, her role is rarely limited to simply conducting research on the end users of a product being created or updated for a project. As the UX Designer, Maggie is nearly always responsible for conducting interviews and site visits with end users, conducting think-aloud protocols with users where she observes users performing tasks with the product and says what they're thinking as they do the tasks, creating personas of different user types from demographic data she collects, creating user experience maps that allow the whole project team to see each "touchpoint" where users interact with a product so that the team can design for those experiences, and much, much more.

However, because the understanding Maggie develops through her research is so thorough, she rarely stops serving a team by merely reporting on the results of her research. She is often the team member who creates the microcopy for the UI since, given her work with protocol analyses, she knows the terms users are seeking when they wish to complete a task. This makes it easy for her to create the labels for buttons and terms in pull-down menus for the interface.

She also understands the mental models that users have when they approach the interface, so she's able to create effective instructions on how to complete specific tasks, and she's able to work with the graphic designer to develop visual metaphors and icons that allow users to utilize the interface more successfully. And of course, when it comes to helping the software engineers break the tasks users need to perform down into manageable parts so that the engineers can write their code in modules that handle discrete bits of functionality in the interface, Maggie's work on user experience maps makes her an invaluable collaborator for the engineers. Bottom line: Every team Maggie supports has different needs, so she needs to bring a far wider range of skills to her job than her job title might suggest.

In this chapter, we'll look at some of the basic tools Maggie uses to help her teams develop "empathy" for their users. These tools include goal-setting interviews, context inquiry, user stories, protocol analyses, eye-tracking, and SWOT analyses. The next chapter will show how the data Maggie collects are used to build more elaborate UX models, such as personas and user journey maps.

What Is Goal Setting with a Client?

Whether you're an independent contractor working for an external client or a member of a UX writing team beginning a new project at your company, chances are that you're going to need to have what's called a "kick-off meeting" with either the leadership team for an external client or the project manager for an internal company. The kick-off meeting can be critical to the success of your project and how smoothly the work goes over the rest of the project lifecycle. The more time you're able to spend on the front end getting your goals outlined and prioritized with your client, the less time you're going to have to spend backtracking and revising later.

When you first meet with a client, it's useful to have them tell you the story of how the need for the product or service came into existence. First of all, asking people to tell you "the story" behind a project is helpful because storytelling is something that comes naturally to most people; it's a way of putting your client at ease. Also, getting the backstory on a product can really help you contextualize the work you're doing.

Another useful strategy is to ask the client to describe five deliverables or outcomes that they would like to see, and then, once they've articulated those five, ask them to pick the top three and put them in rank order from most important to least important. Clients, whether internal or external, nearly always want way more than is possible to produce with the time and resources allocated to a project, and this helps the client understand that they're not going to get everything they want. More importantly, it helps you understand what they consider to be their most desirable outcomes. Finally, during this portion of the meeting, it's important to have the client help you understand how they would measure success for their top three priority outcomes. What are your "successful completion criteria," or SCCs?

Yet another useful strategy during the kick-off meeting is to ask clients to describe what you should *not* do. For example, is there proprietary, mission critical information that needs to be confidential and should not be discussed outside the development team? Are there particular competitors, departments, or individuals who should *not* be invited to participate in the product's development? Are there any intellectual property or privacy issues about which you need to be aware? Knowing about these constraints can help you map your way through the potential minefield you'll have to navigate.

Of course, you'll also need to collect basic information about the product or service being developed. Who are the target users? What's the timetable? What resources are available for you (tools, equipment, personnel, and financial)? Another critical component is who will be your liaison during the development process. Even on an in-house development team, there's usually someone to whom you're supposed to report and with whom you should be consulting about problems you might be having or approvals you might need for different stages in the project's completion process.

During your meeting(s) with the client, you want to have a member of your team who's totally dedicated to taking notes and capturing the information provided because, at the end of the meeting, you'll create a one- to two-page memo, often called a "problem-purpose statement" or a memo of understanding, which details the directives you received and outlines what you understand the goals of the project to be. You'll provide this memo to your client and ask them to correct anything you may have misunderstood or that may have been mistakenly communicated. This document is very important because it provides written documentation in case there's a question later on about whether you've successfully completed the terms of the contract and performed the work you were assigned.

According to 247meeting (https://247meeting.com/kick-off-meeting), the following are some of the key goals and purposes for your kickoff meeting:

- **Formally meet with your client/customer** – This is your team's best opportunity to understand the client's objectives, way of thinking, and expectations. It's important to listen to what they say and what they don't say.

- **Understand the background of a project** – You should provide context as to why this project is relevant and why it's worth undertaking.
- **Understand what success looks like** – The overall goal of the project should be set during the kick-off meeting. Provide a target or goal in the early stages of the project.
- **Establish what needs to be done** – After the kick-off meeting, everyone present should have a complete understanding of what needs to be accomplished in order to achieve success.
- **Create a common understanding of roles** – Your team should have a mutual understanding of each other's responsibilities within the project.
- **Agree on how to work together effectively** – Your team should establish effective means of collaboration to remove any potential barriers to success.
- **Set the tone for the rest of the project** – Start as you mean to go on. Establish the approach for all future meetings.
- **Set expectations** – Illustrate to your team the realistic measures of success to be anticipated along the way.
- **Minimize potential surprises** – Anticipate any issues or unexpected delays that may occur, acknowledge them, and set a contingency plan in the case of each.
- **Size up each other's strengths and weaknesses** – Know where your team is at their best and worst. Plan accordingly and establish means of utilizing each individual's best skills.
- **Generate enthusiasm and inspire confidence in your client/customer** – Ensure that your client/customer leaves the meeting feeling invigorated by the level of commitment and enthusiasm from your team.

What Is Contextual Inquiry and Task Analysis?

The contextual-inquiry method was developed by Hugh Beyer and Karen Holtzblatt in their classic, 1997 book *Contextual Design: Defining Customer-Centered Systems*. Long considered a crucial research method for developing empathy for customers and creating insights into users' "mental models," it is a type of ethnographic study that involves the researcher(s) being embedded for an extended period of time in the actual contexts where the products or services are being used. As Jim Ross (2012) observed, "The key differentiator between contextual inquiry and other user research methods is that contextual inquiry occurs *in context*. It's not simply an interview, and it's not simply an observation. It involves observing people performing their tasks and having them talk about what they are doing while they are doing it." Thus, a contextual inquiry goes well beyond the types of information you can collect in the goal-setting interview described above. First, in goal setting, you're meeting with clients, whereas in contextual inquiry, you're meeting with users. Also, as a researcher, you're not dependent on users' memories or their tendency to ignore factors in their workplaces that they take for granted because they're always there. Because you're onsite with the users, you can *directly* observe what they no longer notice, so you can empathize much more easily with them.

According to a 2010 entry on the Usability Professionals Association's (now UXPA) Usability Body of Knowledge website (see https://www.usabilitybok.org/contextual-inquiry), there are four principles behind contextual inquiry:

1 **Focus** – Plan for the inquiry, based on a clear understanding of your purpose.
2 **Context** – Go to the customer's workplace and watch them do their own work.

3   **Partnership** – Talk to customers about their work and engage them in uncovering unarticulated aspects of it.
4   **Interpretation** – Develop a shared understanding with the customer about the aspects of work that matter.

> The results of contextual inquiry can be used to define requirements, improve a process, learn what is important to users and customers, and just learn more about a new domain to inform future projects.
>
> (*Contextual Inquiry*, 2010, n.p.)

And yet, although contextual inquiries are a powerful tool in the UX writer's toolbox, they're not always possible to perform. Frequently, researchers simply cannot obtain permission to access the users' site. Many companies are extremely sensitive about their "intellectual property" and believe the workplace practices and development processes unique to their companies need to remain secret to maintain a competitive advantage in the market. They don't want outsiders to learn the "secret recipe" to their success, and so management frequently prohibits the access necessary to conduct a contextual inquiry.

Also, some environments are just too dangerous or require too much-specialized training to allow for outside observers. For example, in order to study ways to improve the ways that emergency medical response teams communicate with each other and with hospital personnel for her award-winning book, *Rhetorical Work in Emergency Medical Services*, Elizabeth Angeli (2018) had to undergo a rigorous series of courses and receive certification before she could ride along with emergency medical services (EMS) personnel in ambulances. Given that few research teams have the time and motivation to undergo the training that Angeli did, they need to use alternative research methods such as "user stories" or "protocol analyses," which we'll describe next, to approach the types of empathy that contextual inquiries can provide.

What Are User Stories?

User stories are primarily used by software development teams, particularly on agile and extreme programming software development teams. Essentially, user stories are short, simple descriptions of a software feature told from the perspective of the person (i.e., a user or customer) who desires a specific type of functionality from the software. User stories help teams create a shared vision of what they're building, who they are building it for, and why they're building it. They allow the team to focus on what's called the "minimum viable product" so that they work efficiently and don't waste time building functionality into an interface that's not really needed. In other words, as Jeff Patton (2014), author of *User Story Mapping*, explained, "The minimum viable product is the smallest product release that successfully achieves its desired outcomes" (p.33). User stories allow teams to determine what those desired outcomes are so they can work collaboratively on the different small components or modules of the software to create a viable whole.

User Story Templates

User stories begin by following the same simple template format:

*As a < **type of user** >, I want < **some goal** > so that < **some reason** >.*

For example, let's say a team was developing a software interface that would allow college students to register for their classes. Some of the user stories that might emerge for such a software package might include:

As an **English major**, I want to **register for the Fall semester** so that I can **continue to make progress on my degree**.

As an **academic advisor**, I want to **review what courses one of my advisees needs to take to complete their degree if they change majors** so we can **decide how long their graduation might be delayed**.

As a **student**, I want **to see what courses are being offered next semester** so I can decide which are of interest to me and **which best fit my work schedule**.

As a **student**, I want to **review my degree progress** so I can see **which classes I still need to take to graduate**.

User stories are often written on three-inch by five-inch index cards and then transferred to temporary, usually color-coded sticky notes so that they can be arranged on corkboard walls, whiteboards, or tables where design teams can use them to decide which stories can be combined under one major rubric, which will need to be broken down into smaller sub-routines and tasks, and which will have higher priorities than others. As Figure 4.1 shows, the user stories collected by different members of the course registration design team have been given a task name and transferred to sticky notes so they can be easily moved around. In this case, because the user stories originated from several different members of the design team, they needed to be grouped together under major classifications so the team could see how they were related. Different members of the team created user stories for "Find out how to apply," "See required application documents," "Find a link to the application page," and "See contact info regarding application questions." Because these tasks and

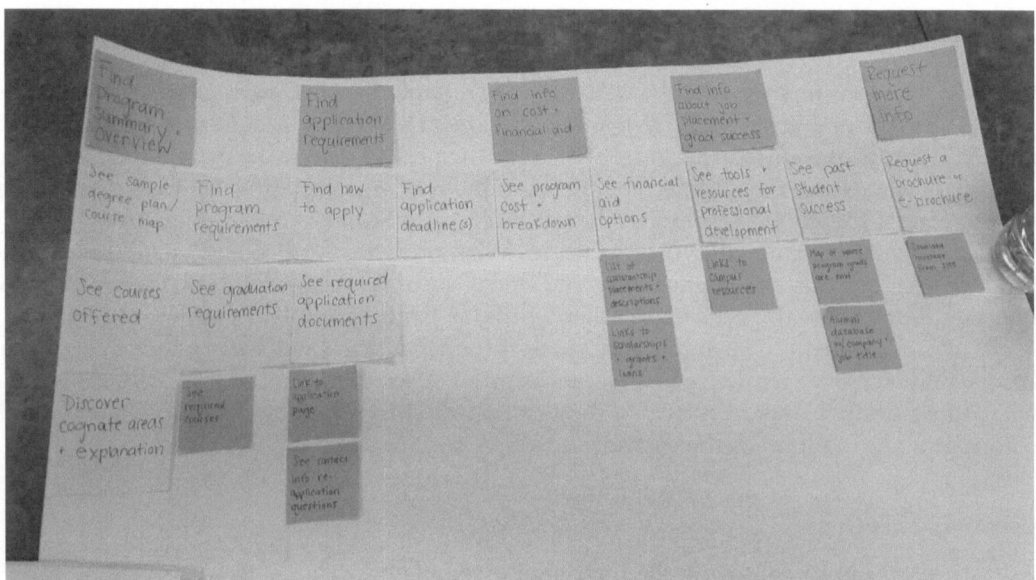

Figure 4.1 Sticky notes arranged into a storyline.

their associated user stories were all related to application requirements, the team decided to group all these together under the major group "Find Application Requirements."

The 3 C's of User Stories: Card, Conversation, and Confirmation

As Figure 4.1 hopefully makes clear, individual user stories aren't really all that useful in and of themselves. A development team will have many user stories, but it's the process of using them that makes them a powerful tool. In 2001, Ron Jeffries is credited with having first described that process when he wrote that "User stories have three critical aspects. We can call these Card, Conversation, and Confirmation" (Jeffries, 2001, n.p.).

Card

Usually called "the 3 C's (Card, Conversation, Confirmation) of user stories," the process begins, as we've already seen, with the creation of the tripartite, formulaic statement of a user story on an index card. The "card" is the index card with the "*As a <type of user>, I want <some goal> so that <some reason>*" statement written on it. However, Jeffries says of cards that:

> User stories are written on cards. The card does not contain all the information that makes up the requirement. Instead, the card has just enough text to identify the requirement, and to remind everyone what the story is. The card is a token representing the requirement. It's used in planning. Notes are written on it, reflecting priority and cost. It's often handed to the programmers when the story is scheduled to be implemented and given back to the customer when the story is complete.
>
> (Jeffries, 2001, n.p.)

Conversation

Cards are really just "grist for the mill" and are intended to be the starting point for *conversation* among members of the design team. In and of themselves, the cards alone don't really provide sufficient information for a UX writer, a designer, or a software engineer to begin building a product. A card and its attendant user story have to be put into the larger framework, or what Jeff Patton (2014) calls the "story map" of the product. Often, for example, a user story will have enormous gaps that need to be filled before it can reach a satisfactory conclusion. There are nearly always several subroutines and preliminary tasks that need to be completed before a user story can be implemented, and Figure 4.2 illustrates a team engaging in a conversation to fill in the gaps.

Sticking with our example of a software development team creating an application that will allow college students to register for classes, Figure 4.2 illustrates how team members add subroutines and tasks that have to be completed before a user story is complete. Under the user story named "Determine needed courses," there are several new cards representing subroutines that the team has identified as needing to be built first. Before the process of determining courses can begin, the user has to login in to the registration system (called "iRoar" at this school), choose the department and curriculum they need, choose the semester in which they wish to register, see what courses are available, and so on. Creating the story map allowed the team to find gaps that needed to be filled to create the minimum viable product.

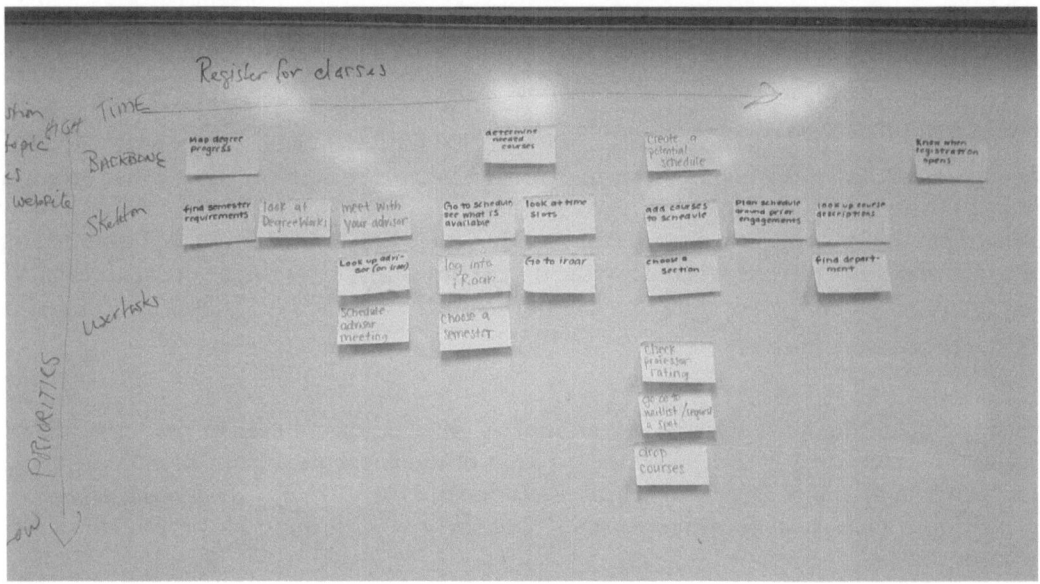

Figure 4.2 Whiteboard conversation about user stories.

The story map also serves the purpose of helping the team decide which functionality should receive the highest priority as they build it. For example, under the card named "Create a potential schedule," there is a list of tasks in descending order of priority. The highest priority functionalities are "Add courses to the schedule" and "Choose a section" of a course. Consequently, these appear higher in the column of the map. Less important are "Checking the professor rating" for a course and "Adding your name to a waitlist" in order to get into a course that is full. Thus, the conversation and the story map that are created as a result help the team create a shared vision of what gaps need to be filled and in what order the application's modules need to be programmed in order to achieve their final goals.

Confirmation

The third phase of the 3 C's process is *confirmation*, and it basically deals with the question, "How do we know when we have successfully met the goals of the user stories?" What are the "SCCs" or successful completion criteria the team has to meet in order to validate the quality of their work? By convention, the SCCs, or as it is sometimes also called, "acceptance criteria," are written on the back of the user story index card. So, on the front of the card, the team will see the "*As a < type of user >, I want < some goal > so that < some reason >*" statement. On the back of the card will appear a list of the criteria to be measured empirically or through some other testing method in order to prove their success. The next two sections will discuss popular methods for completing acceptance testing: (1) Protocol Analyses and (2) Eye Tracking Studies.

What Are Task-Based Usability Tests (aka "Protocol Analyses")?

Task-based usability tests, originally known as "think-aloud protocols," are a form of observational research that can yield tremendous amounts of useful information about how your users interact with a product. In today's workplace, the term "usability testing" is used to cover a wide range of testing techniques and has subsumed the term "think-aloud protocol." In fact, we will be discussing "usability testing" as a research method again in Chapter 7, where we discuss how to test and validate prototypes created during the ideating phase of the design thinking process.

But in this chapter, we're discussing methods of *empathizing* with users, and thus here we're interested in using task-based usability tests to understand users' goals, their expectations, and the mental models and conceptual frameworks they bring to their experience of a product. Consequently, to highlight the different purposes for using task-based usability tests during the empathy phase, we will primarily use the older name "think-aloud protocol" when describing task-based usability tests used to build empathy with users.

Basically, a "think-aloud protocol" is when you ask a representative user to perform a normal task with a product and to say out loud what they are thinking as they perform that task. Say, for example, that you are working with an e-commerce website and want to understand how users use the site to make a purchase decision about specific kinds of products sold on your site. One of the best ways to answer these questions is to use a protocol analysis research study to listen to actual users talking about how they're making purchase decisions in real time on your site. You could observe what kinds of visuals help users make decisions, which types of text users do and do not read, how users scan your site's page layout, and much, much more. Essentially, you could use a usability study to validate whether or not your site is providing the type of user experience you're seeking and a protocol analysis to understand what expectations and conceptual frameworks the user brings to your site.

Unlike surveys, questionnaires, and even interviews, think-aloud protocol analyses give you the opportunity to observe your *real* users actually performing tasks with your software, website, instruction manual, or other product. You can learn a lot from interviews, to be sure, but users can't always tell you what they're really doing with, say, a website, and it isn't until you actually observe them that truly useful data begins to emerge.

Furthermore, think-aloud protocols don't require large numbers of participants in order to give you credible and actionable research findings. According to usability testing guru Jakob Nielsen (2000), "The best results come from testing no more than 5 users and running as many small tests as you can afford." He argues that "As you add more and more users, you learn less and less because you will keep seeing the same things again and again" (https://www.nngroup.com/articles/why-you-only-need-to-test-with-5-users/). Five users are enough to uncover about 85% of usability problems (more on this model in Chapter 7, Figure 7.2), and Nielsen does go on to say that you need at least 15 users to find all of the design flaws in software interfaces. It's also essential to note that Neilson is assuming that your participants are a representative sample of your actual users and that you're working with a fairly simple product design. Still, even though Nielsen's claims are controversial, the fact that you don't need to survey hundreds of people to get extraordinarily useful results from a think-aloud protocol makes this one of the most popular research methods in all of UX research.

Types of Think-Aloud Protocols

Although there are many different variations on the ways protocol analyses can be conducted, they generally fall into one of two different types: (1) Active Intervention Protocols or (2) Natural Environment Protocols.

Active intervention protocols would probably be better named "facilitated" because researchers are careful not to "intervene" in what users are doing in order to avoid biasing their behaviors. Still, as its name suggests, the researchers actively encourage users to keep talking as they complete tasks with the system or product being studied. A member of the research team (known as the "facilitator") may even ask users to pause what they're doing and explain what they're thinking at a particular moment in the process. Often this happens when users get so involved in what they're doing with, say, a website that they stop thinking aloud, and the facilitator will need to prompt them to continue saying what they're thinking by asking a question like, "What are you thinking, right now?" or "How are you feeling, right now?"

Other times, the facilitator may ask questions intended to clarify the data collected. Users will often provide cues as they're working on a task, such as making a face at the screen, grunting with a "humph," or saying something like "that's weird." Because it's difficult to interpret whether these user responses to the product are positive or negative comments or even possibly something entirely different, the facilitator will ask for clarification. For example, they might ask, "What did you mean when you said, 'that's weird'?" "A moment ago, you said 'humph.' Can you explain what prompted you to do that?" Basically, the facilitated protocol format allows research teams to have a conversation with their users as they are actually performing tasks of particular interest to a design team.

Another advantage of these protocols is that they can be conducted in a wide variety of environments and don't require a special laboratory setup. Ideally, the researchers will have some method of capturing the computer monitor or phone screen the user is using, and they'll also need to record the participant's voice as they think-aloud. If possible, it's also useful to have a second camera focused on the user's face in order to capture any smiles, grins, grimaces, scowls, or other facial reactions to the product. It's also useful to have at least two members of the research team onsite during the protocol. One person is the facilitator mentioned previously, and their role is to administer the study. They provide instructions to the users, ask pre- and post-study interview questions, and, of course, prompt the user with questions during the protocol.

The other member of the research team is the "data logger." The data logger is responsible for making sure that the audio and video (A/V) recording equipment is operating properly and is, in fact, recording data. This is more important than many people think because collecting data is the *raison d'etre* for doing the study in the first place. So, if your team is not recording the data, you're wasting time and resources. They are also, as their title implies, responsible for taking notes and logging the data so that it is keyed to the A/V recordings. In other words, the data logger will usually make a note of comments or actions that the user performs and then key that information according to a time code. This will allow the research team to go directly to any point on the recording to review a specific comment or a specific behavior that is of interest without having to re-watch the entire video. They might, for example, be interested in a point during the protocol where a user was struggling to complete a task. The data logger's notes should be sufficiently detailed so that the team can go directly to the point in the recording where the user's struggle occurred.

Unlike active intervention protocols, where the facilitator and the data logger might be in the room with the study's participant and may even be onsite at the user's office, natural environment studies are typically conducted in a laboratory setting such as the one shown in Figure 4.3. Once again, the name "natural environment" is a bit misleading because any study that takes place in a laboratory setting isn't really "natural." However, the point is that the researchers attempt to recreate the typical environment where users are likely to be using the product being studied. In many cases, this would be a traditional office environment like the one shown in Figure 4.3.

As the figure illustrates, there are at least two rooms in the usability testing facility; some labs have a third room for observers (e.g., project managers, engineers, marketing specialists, etc.). The two rooms are soundproofed and separated by a one-way mirror so that the user who is completing the study's tasks in the "Observation Room" will not be distracted by the facilitator, data logger, or observers. Again, as we observed in contextual inquiries, it's not always possible to observe users in their actual work environment, so the goal in the observation room is to create a "natural" environment such as a business or home office where the product being tested (usually software or a website) is normally used.

The facilitator has the ability to communicate with the user via a microphone in the control room and speakers in the observation room; however, unlike active intervention protocols, the goal is to disrupt the user as little as possible. One important reason for this is that in the natural environment, protocols often measure the user's "time on task." Asking a user to stop and explain a comment they just made corrupts data like time on task, potentially biasing the results of the study. The user's think-aloud comments are recorded and coded into behavioral categories, so it is important that the facilitator prompt the user to keep saying what they're thinking and doing as they perform the tasks. Indeed, often the research team will establish parameters for the study that determines how long a study

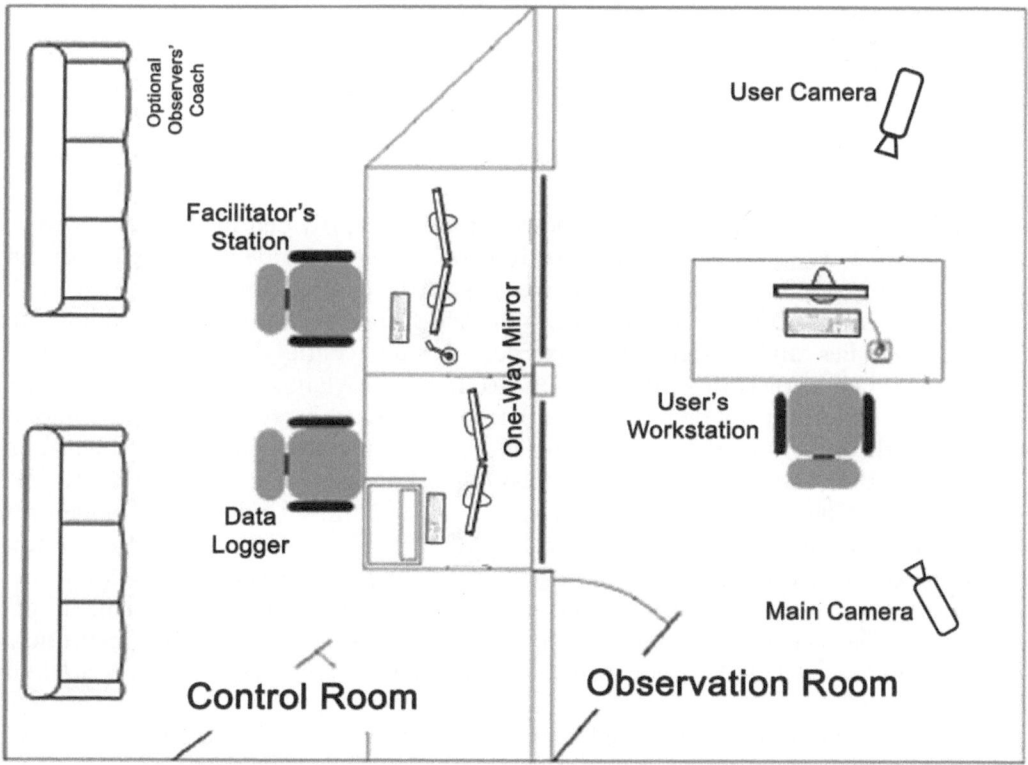

Figure 4.3 Floorplan of typical usability testing facility.

participant can go without speaking. For example, the team may decide that the facilitator will need to prompt a participant if they haven't spoken for five seconds, ten seconds, or another predetermined time period.

Parts of the Think-Aloud Protocol Studies

Regardless of the type of think-aloud study being performed, it typically has four parts:

1 Informed Consent
2 Pre-Study Questions and Interview
3 Think-Aloud Protocol
4 Post-Study Questions and Interview

Informed Consent

The informed consent portion of the study is where the research team obtains the study participant's permission to record their protocols, answers to interview questions, and other data. And before they give their consent, the participants in the study must at least be informed of:

- the purpose of the research,
- any risks or discomforts associated with the research,
- any potential benefits of the research,
- how an individual's confidentiality will be protected,
- how much (if any) compensation they will receive for participating,
- how to withdraw their consent,
- contact information if they have questions or concerns about the study.

These are the basic types of information that must be provided in the U.S.; however, most countries have laws that govern how research studies may be conducted with their citizens, so you should familiarize your team with those government policies if you're working with international products. But in the US, there are important federal laws that govern how studies can be approved and conducted at universities. Universities have something like an "Office for Human Research Protections," or what is more frequently known as an "Institutional Review Board" (IRB). Before any research project where the findings will be published or presented at a conference can be conducted by faculty, students, and staff, the research team must obtain permission to conduct the research from the IRB before recruiting participants and beginning the study. Failure to do so is a violation of federal law, so you should use caution here. Even though many IRBs exempt research studies that are conducted in classroom settings because the research results aren't published, other institutions may not. Furthermore, many institutions require that members of a research team be "certified" through online courses such as the Collaborative Institutional Training Initiative (CITI) Program before studies may be conducted (see https://about.citiprogram.org/).

Below is an example of an informed consent statement approved by the IRB at Clemson University to conduct a study on the ways that college students use grammar handbooks in their writing classes.

Consent Form for Participation in a Research Study

Clemson University Usability Testing Facility
(Usability Testing Educational Handbooks and Materials)

Description of the Research and your Participation

You are invited to participate in a research study conducted by the Clemson University Usability Testing Facility under the direction of Dr. Tharon Howard. This study is sponsored by Allyn & Bacon/ Longman Publishers, and the purpose of this research is to make textbooks easier to learn and use.

Your participation will mainly involve answering survey and interview questions. We will also ask you to use the textbooks provided in ways that they would normally be used in an educational environment, and we will ask you to say what you are thinking aloud as you use the text. Once we have your permission to do so, we will begin videotaping your responses to questions and to the textbooks.

The amount of time required for your participation will be approximately two hours.

Risks and Discomforts

There are certain risks or discomforts associated with this research. They include the possibility that an excerpt of you on video may be seen in a public setting.

Potential Benefits

By participating in this study, you will help us make future textbooks and other educational materials easier for you and other students to use and understand.

Protection of Confidentiality

Once we have your permission to do so, we will begin videotaping your responses to questions and your use of the textbooks. We will use the videos you give us, along with the information we collect from other people, to recommend ways to revise the textbook being studied and to develop general guidelines for creating useful textbooks and educational materials. We will be taking excerpts of comments that you and other participants make and will be compiling these into a "highlight" video, which we will use to present our findings. This highlight video will be used by the authors and publisher of the textbooks to improve the work being studied as well as that of other texts. However, your name will not appear in any of the video clips, and we will do everything we can to protect your privacy.

Federal regulations require that we keep our data for a minimum of three years; however, our "raw" videotapes, notes, and other materials will be kept in a locked cabinet in the Usability Testing Facility, where they are only accessible to members of the Usability Testing Facility staff. The tapes will be destroyed after three years. Also, our highlight videos will not use clips in which participants provide information such as locations, addresses, actual names, or demographic data that could be used to identify them. Your name will not be revealed in any publication that might result from this study.

In rare cases, a research study will be evaluated by an oversight agency, such as the Clemson University IRB or the federal Office for Human Research Protections, which would require that we share the information we collect from you. If this happens, the information would only be used to determine if we conducted this study properly and adequately protected your rights as a participant.

Voluntary Participation

Your participation in this research study is voluntary. You may choose not to participate, and you may withdraw your consent to participate at any time. You will not be penalized in any way should you decide not to participate in or withdraw from this study.

Compensation

For satisfactory completion of the study, you will be paid $75.00.

Contact Information

If you have any questions or concerns about this study or if any problems arise, please contact Dr. Tharon Howard at Clemson University at 864.656.xxxx. If you have any questions or concerns about your rights as a research participant, please contact the Clemson University IRB at 864.656.####.

Consent

I have read this consent form and have been given the opportunity to ask questions. I give my consent to participate in this study.

Participant's signature:　　　　　　　　　　　　　　　　Date:

A copy of this consent form should be given to you.

Pre-Study Questions and Interview

The pre-study question and interview section of a study is where your research team has an opportunity to collect data on the basic demographic information of your users, such as age, education level, gender, and so on. But more importantly, it's also where you can collect information about the user that might potentially change the way the user completes the tasks.

For example, in the case of the study of grammar handbooks, where the sample informed consent statement was used, one of the research team's concerns was that we didn't have participants who were all exemplary students or who were already so familiar with topics like how to create an MLA citation for a research paper that they wouldn't use grammar handbooks in the first place. As a result, the questions we used asked the study participants to discuss their previous experiences writing papers in high school and any college courses they may have taken. We also told them that we were looking for "typical, average students" rather than above-average or below-average students, and then (having explained why we were asking about the sensitive subject of grades), we asked them to talk about their performance in English, History, or other courses where they may have been required to write.

During this phase of the study, there are two key points to keep in mind. First, as you may have noticed already, we have *not* used the terms "test subjects" or "pre-test questions" to refer to the users or participants in a "study." There are several reasons for this, but the main one is that nobody likes to be tested, and nobody likes to be subjected to anything. It's important that your study's participants feel comfortable enough to be honest and open with you (particularly once you ask them to start "thinking aloud"), so you want to avoid calling your study a "test" or using terms like "quiz" and "evaluation" when you work with your study participants.

The second thing to achieve during the pre-study interview phase is to *keep it short*. Users have limited patience and tire easily during a study, and a common mistake many first-time researchers make is to ask far, far too many questions during the preliminary phase of a study. Basically, the members of your research team need to examine each question you're asking and then ruthlessly chop any questions that are not absolutely essential to the findings or that duplicate information you can collect through other means.

Think-Aloud Protocol

This part of the study is, of course, the main, mission-critical portion of the study. This is where you will ask the user to think-aloud as they perform tasks, and there are typically at least three parts to this phase of the overall study: (1) the instructions, (2) the scenario or scenarios, and (3) the tasks.

The sample document below provides an example of the instructions, scenario, and task used in the grammar handbook study. The instructions, as the document illustrates, are a critical component of protocol analyses because it's essential that users understand that *they are not being tested or evaluated*—instead, it's the product they're using that is being tested. Users need to understand that it's absolutely okay for them to be critical and negative about the experiences they're having with the product. There's a tendency for people to hide their harsh and critical responses to an experience when they're in a public setting, so you have to make explicit to them that what you're looking for is precisely that type of information so that you can revise and/or rebuild the product to ensure that future users don't have the same negative experiences. In fact, one really good technique for getting this point across to users is to model a "mini think-aloud" for them. For example, the study facilitator can take a short paragraph from a set of instructions on, say, how to use their cell phone, and then they can say what they're thinking and make *both* negative and positive comments about the experience as they're doing it in order to illustrate for users what it is that you're seeking from them.

The second part of this phase of the study is the scenario. In this example study of grammar handbooks, we wanted the users to imagine that they were taking an introductory first-year writing class because that's the class where the handbook publisher expected to sell their books. Thus, the scenario created the appropriate "context of use" for the task we were interested in studying.

The third part of this phase is the tasks you ask your participants to perform. In the case of the grammar handbook, there are quite literally hundreds of tasks we could have asked our users to perform. We might have asked them, for example, to define what a "comma splice" entails. However, you only have your participants in a study for about an hour—if you ask users to think-aloud for longer than an hour, they become so fatigued that your data becomes corrupted. So, you need to prioritize the tasks you want users to perform. In the grammar handbook case, our marketing research on why writing teachers adopt grammar handbooks showed that teaching students how to create accurate MLA and/or APA citations was one of the top reasons teachers had their students purchase grammar handbooks, so the task of citing sources was our top task for our users.

Preliminary "Think-Aloud" Instructions

I would like to thank you again for agreeing to participate in this study. This room is the usability testing facility, and we'll be here throughout the time we are together. If at any time you wish to take a break, go to the restroom, or leave, then please just let us know. We encourage you to be as comfortable as possible during this experience.

As you may have figured out from our ad, we are in the process of studying how people use writing handbooks. The textbooks that we're going to be using today are low-fidelity prototypes, meaning that they are drafts that are in their early stages of development. With your help, I hope to offer suggestions to people who write, edit, and publish textbooks.

Do You Have Any Questions?

It is imperative that you understand that I am NOT testing you. In fact, you are aiding me in studying the handbooks. There are no right or wrong answers in this study. Because I am trying to understand what works well and what doesn't work well in the handbooks, it is important that you voice your thoughts as you try to perform these tasks. Harsh language or gestures do not offend me; in fact, they can offer a great deal of information about your experience. All that I request is that you speak out loud about what you are thinking and feeling as you are using the texts.

As the study progresses, I may occasionally ask questions about what you are doing and thinking. In return, you may ask me any questions before, during, or after the study. I will try my best to answer your questions; however, be aware that I may be unable to answer some questions since they could possibly influence your opinions and perceptions of the handbooks. I will answer these questions at the end of the study.

Do You Have Any Questions?

Part One: Citing Sources
Instructions:
In this portion of the study, we want you to assume that you are working in your composition class on a research paper that you have been assigned to write for a grade in the course. For the purposes of this scenario, please assume that you've already conducted your research and written your paper. We will give you books with passages marked in them that we want you to assume you have quoted in the research paper you have written.

Your teacher has told you that you need to use the "MLA style" for documenting your sources. Since you've already written your paper and quoted the sources in it, your task at this point is to prepare the "Works Cited" portion of the paper. In other words, we want you to cite the works we will provide you in the correct format.

Do You Have Any Questions So Far?

We will be asking you to use the two handbooks to cite the sources. Before we start, I will tell you which handbook to use for each citation. As you're using the handbooks, it's very important that you talk out loud and say what you're reading on the page and, more importantly, what you're thinking about as you use the handbook.

Sources Part One:

Please write the citation for the first work cited in the space below. Please write your response in legible print, and please be sure to underline where appropriate and to use the specified punctuation:

Post-Study Questions and Interview

The post-study interview is often where you get the most useful information about the users' experiences with a product. They'd used it at that point, and you observed where they struggled with it and where they were pleased with it, so the post-study allows you to drill down and ask the user to reflect critically on their specific experiences. This is also the time when you can ask questions that might have biased the users if you had asked them before or during the study.

Reporting Findings from a Think-Aloud Study

Because the data collected in a think-aloud protocol are primarily video clips and audio recordings of users performing tasks, the recommendation reports that come out of these usability studies don't really lend themselves well to paper reports. Instead, most industry professionals provide their reports in what are referred to as "highlight videos," which might be accompanied by a short, written summary of the recommended changes to the product.

The name "highlight video" is a little bit misleading for some individuals because they're expecting something like highlights from a theatrical movie—i.e., a movie trailer. Instead, what the highlight video provides is a series of video clips of users' comments and behaviors clustered around patterns researchers found in the data. The researchers "code" the users' behaviors and identify repeated comments into patterns. For example, users from a study may find a specific page on a website confusing and disorienting. The highlight video will provide a title screen giving a name to the problem observed and then video clip excerpts from two or three of the think-aloud protocols that show examples of what the users experienced on the problematic pages. Typically limited to about 20 minutes in length, the project managers, designers, engineers, and marketers from a product development team can watch the highlight video rather than read a written recommendation report, and they can obtain a much greater understanding of the experiences users are having with the product than they can obtain from even the best-printed prose report. Or, to put this another way, a development team is far more likely to implement the design changes recommended by the

UX research team when a highlight video is used to show the designers what real users are saying about and doing with their product designs.

How Does Eye-Tracking Work?

People tend to think that when we look at a photo, our eyes take in the entire image all at once. We tend to assume that vision is holistic. In reality, however, our eyes only clearly see an area about 2.5 degrees wide. In other words, if you stretch out your arm and hold up your thumb, the size of your thumbnail is about 2.5 degrees. That small 2.5 degree is called the "foveal system," and it's the only part of your eye that has 100% visual acuity; the rest of your vision is called the "peripheral system," and it provides you with a general, holistic impression of your environment. The peripheral system is really quite good at picking up motion in low-light conditions, for example, but to see something clearly, your eye has to move the foveal system across the object you're trying to observe. This is illustrated in Figure 4.4.

Our eyes are the fastest muscles in our bodies, and so while we aren't usually conscious of it, our eyes are constantly moving from point to point taking in images of the world around us and piecing the images together in our brains. In the figure above, each moment of visual clarity inside the circle is called a "fixation," and a fixation is when the eye-movement pauses briefly to acquire content. Typically, fixations last between 60 milliseconds and 300 milliseconds. Just to give you a sense of how fast that is, a ruby-throated hummingbird beats its wings about 53 beats per second which is about 19 milliseconds. In the example above, there are fixations on the boys' faces, on their hands, on the sock and on the soccer ball. Eye-tracking software typically indicates longer fixations by increasing the size of the circles.

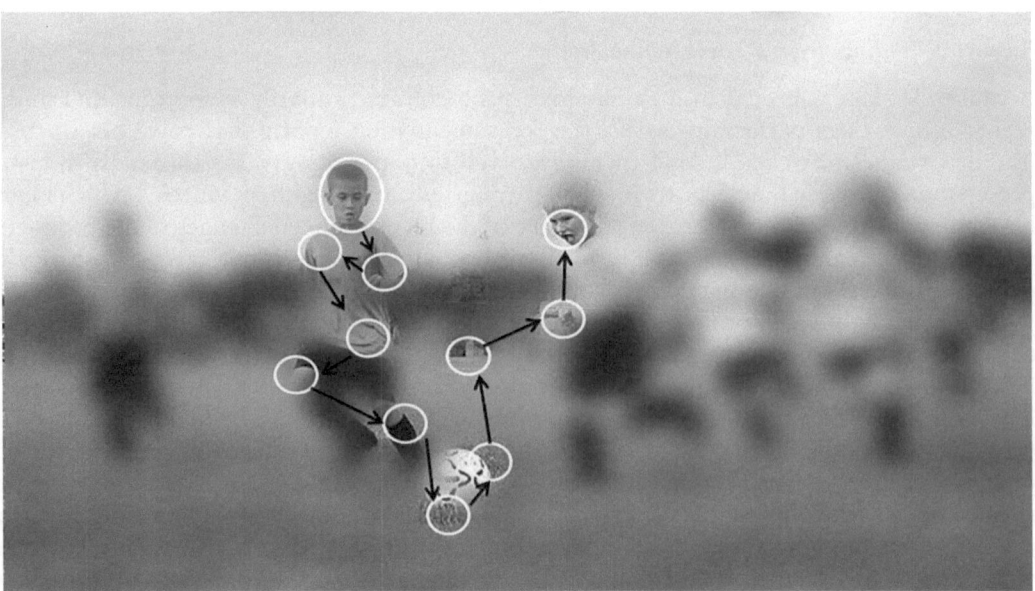

Figure 4.4 Saccades and fixations when looking at a photo of kids playing soccer.
Source: Daniel Liddle, Clemson UTF.

In between the fixations are what are called "saccades," and these are represented by the arrows in Figure 4.4. Saccades are the period of eye movement from one fixation to the next, and they're even faster than fixations. Typically, they range between 20 milliseconds and 200 milliseconds. When you're reading lines of text (like this textbook), your saccades are usually between 20 and 40 milliseconds (which is almost as fast as hummingbird wings at 19 milliseconds).

Eye-trackers take advantage of the fact that our eyes only have a 2.5-degree field of view, and they track our pupils in order to determine where we are looking. There are several types of systems for tracking pupils, but the two major ones are Limbus tracking and Dual Purkinje (a.k.a. "glint"). Figure 4.5 shows a head-mounted Limbus eye tracker that uses a single infrared camera pointed at the eye to track the pupil. Because it's infrared, the user doesn't "see" the LED light being used to illuminate the eye, but the camera is able to collect an image of the eye and more specifically the "limbus" region surrounds the pupil. The "limbus" is the part of the eye that gives it its brown, green, blue, or other color. Software known as "starburst" software is then used to send out rays or lines that find the edges between the pupil and the limbus. As the left side of Figure 4.5 shows, these rays create a "starburst" on the pupil, thereby allowing it to be tracked.

Dual Purkinje can be head-mounted or they can be mounted underneath a computer monitor. These systems are often called "glint" systems because they use an infrared beam of light which is shined into the eye in order to create "glints" or reflections in different parts of the eye. Again, a camera is then used to record the glints, software calculates the angles based on at least two reflections from different parts of the eye (which is why it's called a "dual" system). Once the angles have been calculated, the user's line of sight can then be tracked.

One common misconception about saccades and fixations is that they are random. In reality, however, a very famous study published by Alfred Yarbus in 1967 proved conclusively that saccades and fixations are decidedly goal-oriented and are heavily influenced by the task users are seeking to perform. Yarbus asked participants in his study to examine a painting called *The Unexpected Visitor* shown in Figure 4.6.

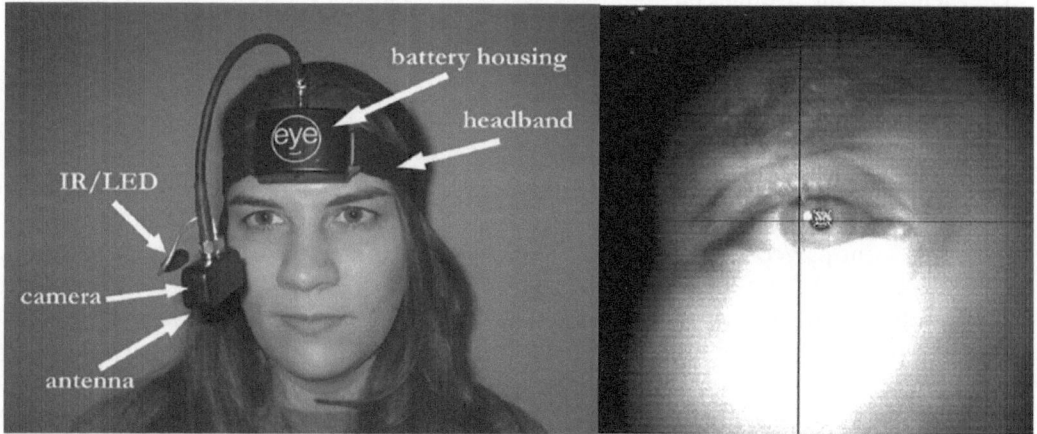

Figure 4.5 Limbus or pupil tracking system.

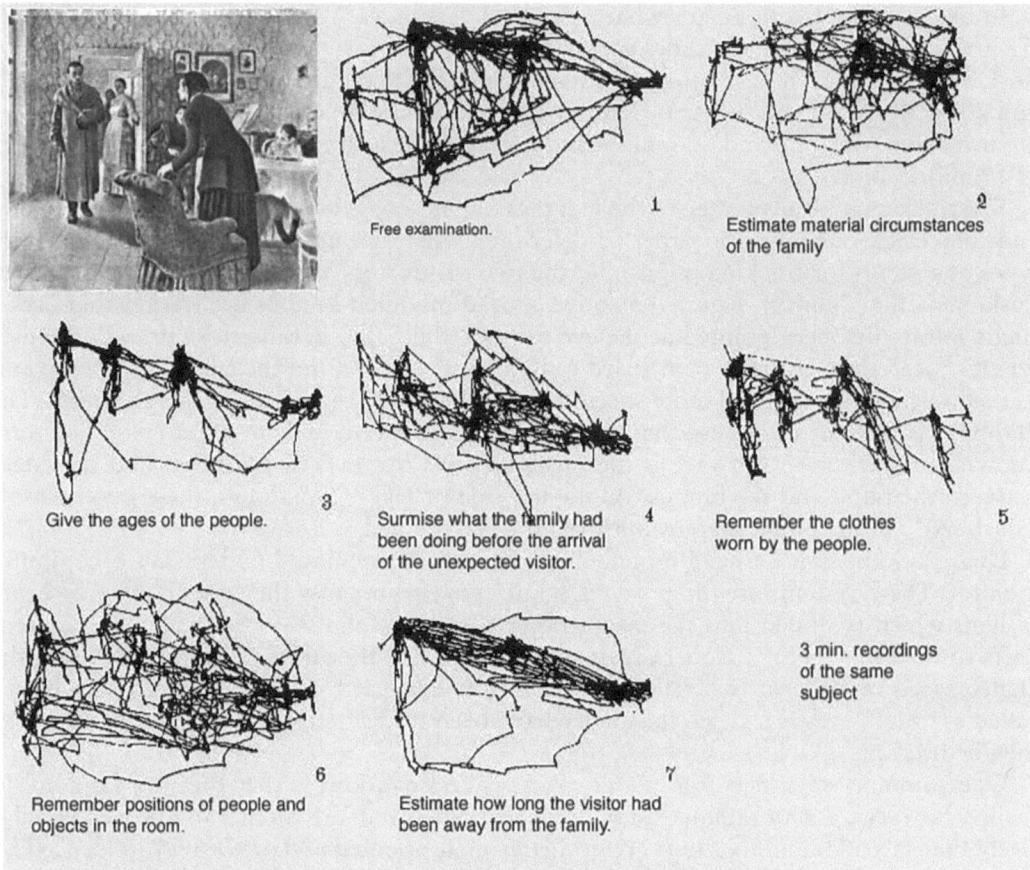

Free examination.   1

Estimate material circumstances   2
of the family

Give the ages of the people.   3

Surmise what the family had   4
been doing before the arrival
of the unexpected visitor.

Remember the clothes   5
worn by the people.

Remember positions of people and   6
objects in the room.

Estimate how long the visitor had   7
been away from the family.

3 min. recordings
of the same
subject

Figure 4.6 Saccades and fixations on a painting from 1967 study by Yarbus (https://commons.wiki-media.org/wiki/File:Yarbus_The_Visitor.jpg#file).

He then gave them different tasks and asked different questions as they examined the image:

(1)   Free examination. (2) Estimate the material circumstances of the family in the picture. (3) Give the ages of the people. (4) Surmise what the family had been doing before the arrival of the "unexpected visitor." (5) Remember the clothes worn by the people. (6) Remember the position of the people and objects in the room. (7) Estimate how long the unexpected visitor had been away from the family.
(https://commons.wikimedia.org/wiki/File:Yarbus_The_Visitor.jpg#file, n.p.)

As Figure 4.6 illustrates, the eye tracks for "3," where users were asked to give the ages of the people in the picture, differ significantly from "6," where users were asked to remember the positions of people and objects in the room. It seems likely that in "3," the users were focusing much more closely on people's faces so they could judge the ages of the people in the image, and users paid less attention to the chair, pictures on the wall, and other objects.

The recognition that saccades and fixations are goal-oriented and task-driven is critical for anyone conducting eye-tracking research studies because it means that the prompts and instructions researchers give to users in their studies can significantly bias the outcomes of their studies. Use care when designing tasks for users to perform during your studies and be sure to conduct pilot tests with actual users in order to ensure that you're not accidentally biasing your research.

Gaze Plots

One of the most popular methods for reporting the results of eye-tracking studies is known as gaze plots. Gaze plots show both the saccades and fixations during an episode with a single user. As Figure 4.7 shows, the circles represent fixations and the numbers inside the circles represent the order in which the fixations happened. This allows researchers to "plot" where the user's gaze went over the course of the episode. The size of the circles is also important because those represent the length of the fixation. The longer the fixation, the larger the circle.

It's worth keeping in mind when you are examining the results of a gaze plot (or any other eye-tracking data) that what the user "sees" is often much more than what the gaze plot suggests. There are at least two reasons why this is the case. The first is the limitations of computing technologies and the speed required to keep up with the human eye. Saccades and fixations can occur so rapidly that computer processes simply can't keep up and the data aren't reported. But the second, and more common issue, is that the eye tracker is only recording what the foveal vision sees. It doesn't record the peripheral vision, which is far greater in size.

Figure 4.7 is an illustration of how gaze plots can be used to see what parts of an image a user is looking at in order to make a decision or evaluation of something. In this particular study, the researchers wanted to discover which anatomical features of typefaces (e.g., bowls, serifs, x-heights, descenders, etc.) did users focus on when asked to decide if a font was "friendly" or "formal" in tone.

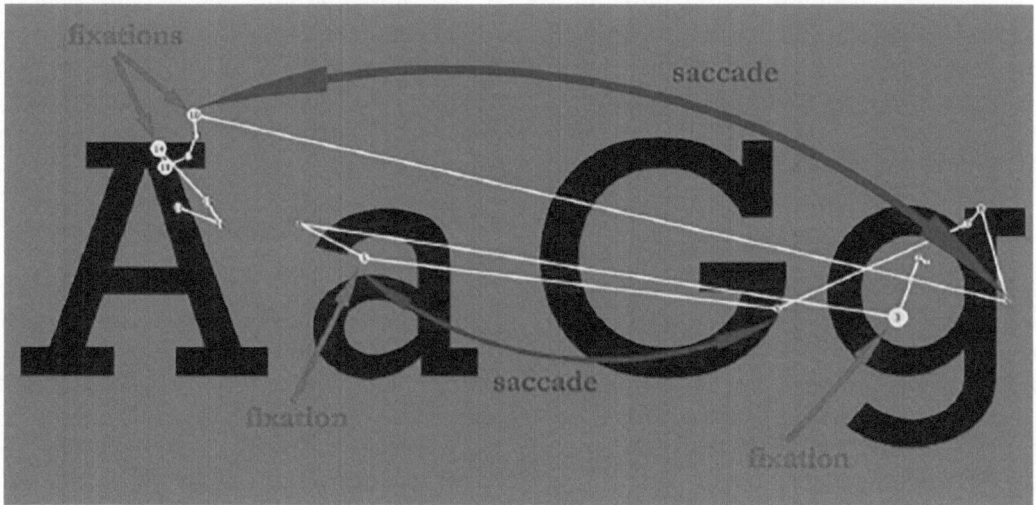

Figure 4.7 Sample gaze plot from an eye-tracking study.

As the gaze plot shows, no fixations occurred for any significant period of time (which is why the circles are all very small), but users consistently looked at the bowls of the letter "g" and the thick-to-thin ratio of the stroke in the letter "A" in order to make their decisions.

Thus, while gaze plots can provide you with tremendously valuable information about what people look at first, second, third, etc., on a screen and can provide information about how long they fixate at various points on an image, it only provides a slice of the information and it doesn't allow you to understand why some fixations are longer than others or why they follow the order they do.

Heat Maps

Gaze plots are useful when you want to look at a single individual's experience with an object or image, but most of the time you want to aggregate your data so that you're getting a sense of where a population of users tends to fixate on objects. Heat maps are one of the most popular ways of providing this kind of information. They look like weather reports showing you different temperatures on a geographical map, which is why they're called "heat maps." However, instead of reporting the temperature where users look, a heat map is a recording of the frequency of fixations in a particular area. Usually, the highest concentration of fixations produces the color red, followed by orange, yellow, green, and blue (with blue having the lowest concentration of fixations in a population).

Figure 4.8 is a heat map of a traditional web page, illustrating what is often called the "F-pattern." With a traditional web page design users typically will scan the major headings on the page and, unless they're interested in the information "contained" within the paragraph, they skip down to the next header. This behavior, which you can see illustrated in Figure 4.8 produces patterns that tend to resemble a capital F or a capital E. Another phenomenon frequently observed on web pages like the ones illustrated below is called the "Golden Triangle." Users all tend to focus on the upper left-hand corner of the screen and then move down to the right; however, the further down they go to the right and toward the bottom of the screen, the fewer concentration of fixations typically occur. Advertisers use this phenomenon to charge more money for "screen real estate," which appears in the upper left-hand portion of the screen. Cheaper ads occupy the less expensive screen real estate on the right-hand side of the page and further down.

Again, like any other eye tracking data, caution needs to be used with the interpretation of the data. As we saw with the gaze plots the trackers don't record peripheral vision and extremely fast saccades and fixations.

Areas of Interest

Like heat, areas of interest (better known as "AOIs") are another useful method for showing aggregated data from the eye-tracks of multiple users. They're called AOIs because researchers have the ability to plot out in advance areas on the screen or image in which they are most interested in collecting data. Figure 4.9 below illustrates this. The top half of the figure shows the "Find a COVID-19 vaccine" website without the AOIs overlay. In this image, you can see that the designers of the website wanted users to be able to find the button where they could click and go directly to information about their state and county so that they could get a map of where to obtain vaccinations in their local area. The researchers also wanted the users to know that the information they were getting was

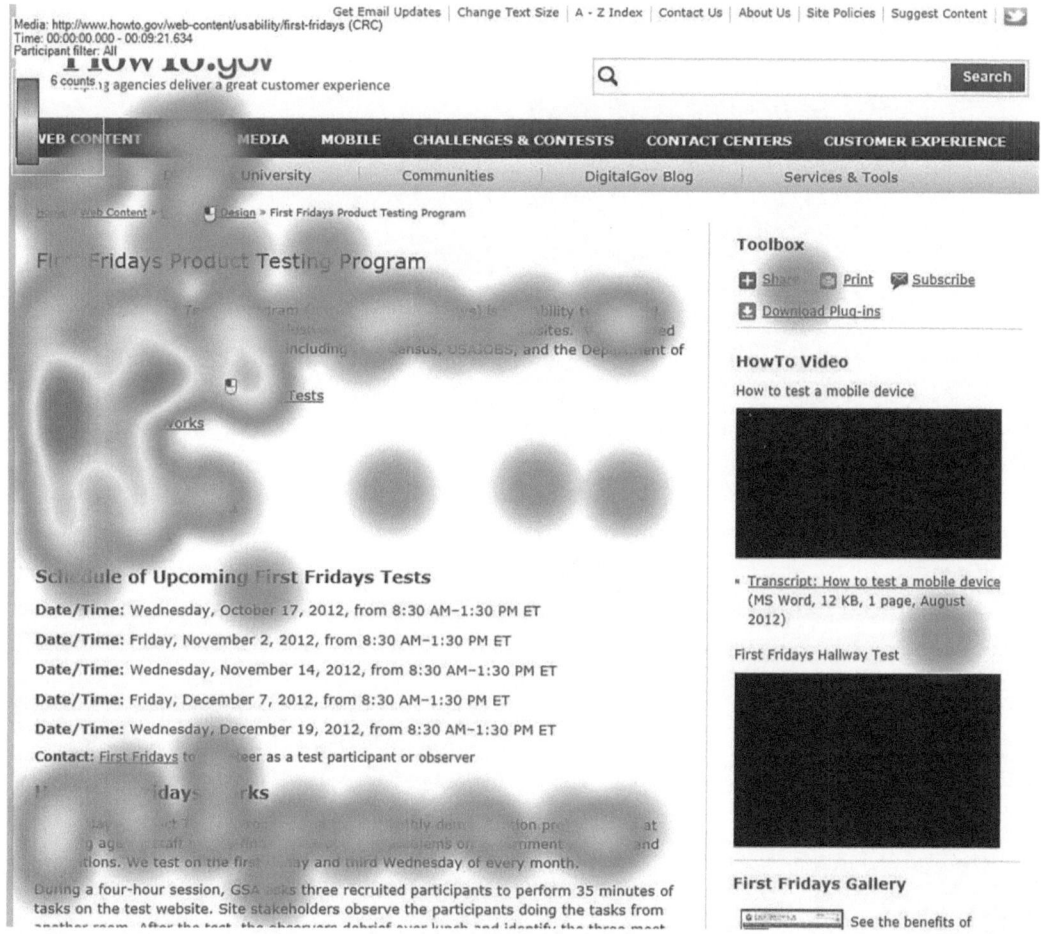

Figure 4.8 Sample heat map of a website homepage.

Source: Digital.gov (https://digital.gov/2014/04/04/heatmapping-tools-show-whats-hot-on-your-pages/)

authorized by the Center for Disease Control so that they would consider it authentic and trustworthy.

As the bottom half of Figure 4.9 shows, the AOIs have a box around them, and inside the overlay displays the total percentage of fixations that occurred inside each box. Thus, the most critical information trying to be communicated on the page received 56% of the fixations. This high percentage of fixations is an affirmation and validation of that aspect of the page's design. However, the fact that the area with the Center for Disease Control and Prevention (CDC) logo only received 13% of the fixations while the image of the woman and child received 29% of the fixations suggests that altering the design might be desirable in order to increase the number of fixations on the CDC logo. Research has shown that humans tend to respond to other people's faces, so changing the image from a young girl to a vial of medicine or some other inanimate object might decrease the percentage of fixations in that particular AOI and increase it elsewhere.

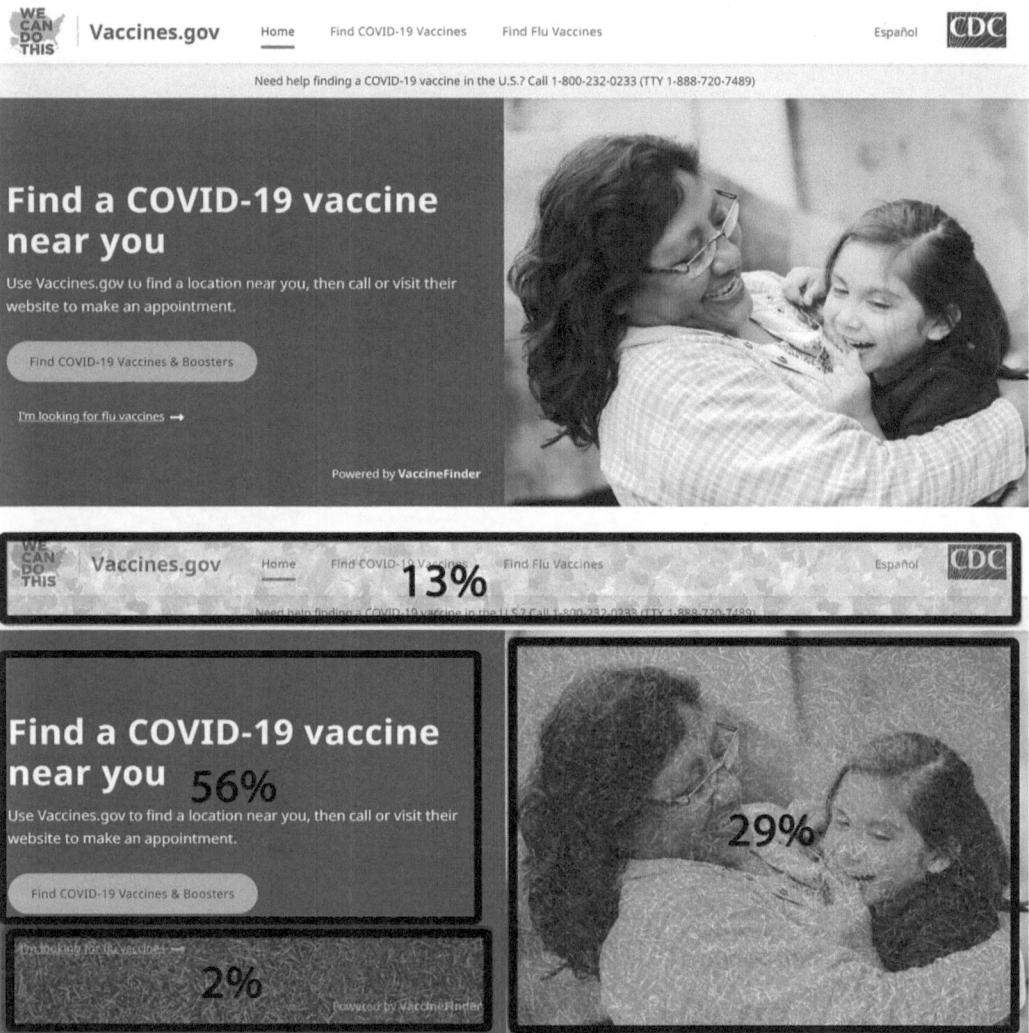

Figure 4.9 Areas of interest overlay on https://www.vaccines.gov/ homepage.

Ultimately, eye-tracking research studies are an exciting and powerful way of gaining insights into what users are seeing as they experience products. Unlike think-aloud protocols, where users will tell you *why* they are looking at a particular graphic on a website or why they read a particular passage on the screen, eye-tracking studies can collect data on where users are looking that they don't always verbalize in their think-aloud protocols. Taken together, eye tracking combined with think-aloud protocols provide UX researchers with tremendous amounts of data on users' experiences.

What Are SWOT Analyses?

SWOT analyses are typically used to examine entire companies, organizations, and/or departments. A SWOT analysis can give you a past, present, and future snapshot of an organization and helps you see what has been working well for an organization, what has been problematic, and what future possibilities—both positive and negative—exist.

SWOT analyses are typically reported in the form of a grid such as the one shown in Table 4.1, and the columns of the grid reflect internal forces on an organization (i.e., strengths and weaknesses) and external forces (i.e., opportunities and threats). The data displayed in each cell of the SWOT table can come from a variety of sources. They can come from data collected from surveys, interviews, and questionnaires, but they are more typically generated by assembling key stakeholders in an organization together and then having a trained facilitator run focus group sessions. Noah Parsons (2021) offered the following suggestions for running your focus groups:

1 **Gather the right people**

Gather people from different parts of your company and make sure that you have representatives from every department and team. You'll find that different groups within your company will have entirely different perspectives that will be critical to making your SWOT analysis successful.

2 **Throw your ideas at the wall**

Doing a SWOT analysis is similar to brainstorming meetings with right and wrong ways to run them. I suggest giving everyone a pad of sticky notes and having everyone quietly generate ideas on their own to start things off. This prevents groupthink and ensures that all voices are heard.

After five to ten minutes of private brainstorming, put all the sticky notes up on the wall and group similar ideas together. Allow anyone to add additional notes at this point if someone else's idea sparks a new thought.

3 **Rank the ideas**

Once all of the ideas are organized, it's time to rank them. I like using a voting system where everyone gets five or ten "votes" that they can distribute in any way they like. Sticky dots in different colors are useful for this portion of the exercise.

Based on the voting exercise, you should have a prioritized list of ideas. Of course, the list is now up for discussion and debate, and someone in the room should be able to make the final call on the priority. This is usually the CEO's job, but it could be delegated to someone else in charge of business strategy.

You'll want to follow this process of generating ideas for each of the four quadrants of your SWOT analysis: Strengths, Weaknesses, Opportunities, and Threats.

(Parsons 2021, n.p.)

The SWOT analysis shown in Table 4.1 is from a graduate student seminar on a content strategy where students in the course were working with a local breast cancer survivor group in order to help them create a content strategy for their group.

In this particular case, the SWOT analysis began with a 2.5-hour-long focus group meeting between five of the leaders of the organization. The information collected was subsequently verified with interviews and surveys of all of the organization's members. However, during the focus group with key leaders, a clear snapshot of the group emerged. The students learned that, although the organization had been around for over 30 years, it had never had any formal organizational structure. It had no "president," bylaws, or officers. Indeed, if someone newly diagnosed with breast cancer wanted to know whom to contact to

Table 4.1 Sample SWOT analysis of breast cancer survivor group

Internal forces	*External forces*
Strengths	*Opportunities*
Motivated membership	Pro-bono lawyer willing to file 503C paperwork
Long-time organization (over 30 years)	Create flyers for healthcare lobbies
Community support	Develop website for calendars, communication, and recruiting
Support from local healthcare providers	Local university students can create logo, brand identification, and content strategy
Social & emotional ties between members	
Weaknesses	*Threats*
No bylaws, governance, or decision-making systems	Cannot accept grants, gifts, or donated resources without 503C legal status
Flagging morale in volunteer leadership	Recruiting new members is totally word-of-mouth
Poor communication channels between members	No mechanisms for communication with potential members or sponsors outside the organization
Dependence on volunteers	

learn more about the organization, there was no one clear person in charge available. This made it extremely difficult for the organization to recruit new members and provide support services to newly diagnosed cancer patients. Furthermore, many businesses in the community frequently offered to donate resources and services such as free therapeutic massages, hair replacement rentals, exercise classes and equipment, and even outright cash grants and donations. However, the fact that the organization lacked legal status as a non-profit (which is known as "503C") meant that neither businesses nor individuals could make donations to the organization.

Conducting the SWOT analysis gave students a useful framework for collecting the data they needed to understand the organization's goals, and it allowed everyone involved to see at a glance where the organization's priorities needed to be placed.

Conclusion

In this chapter, we have presented six qualitative, quantitative, attitudinal, and behavioral research methods that support writing and design in UX writing. Since we've "drilled down" into the nitty-gritty business of how to conduct these types of studies, it's important to remember that our overarching goal in this chapter has been to situate UX writing in design thinking philosophy and user-centered design processes, particularly through the "empathizing" phase of design thinking. We've provided you with methods of observing and empathizing with everyday users through six specific techniques: (1) goal setting interviews, (2) context inquiry observations, (3) user stories profiles, (4) protocol analyses (task-based usability studies), (5) eye-tracking studies, and (6) SWOT analyses. Depending on the resources and needs of your client and the project on which you're working, you will likely use some if not all of these six methods to build a critical understanding of who your users are, what kinds of content they need to achieve their goals, and how best to deliver that content to them. Indeed, the data you collect during this phase of the UX writing and design process provides the essential foundation on which you will build the personas, journey maps, and content strategies we will discuss in the next chapter.

Chapter Checklist

To succeed in UX writing, it's essential to begin by empathizing with your users so that you can become an advocate for their needs. In this chapter, we have zeroed in on six distinct research tools that you can use to create a deep and complex understanding of your users' goals and needs.

- *Set project goals with your clients and prioritize them.*

  - Use storytelling to have the client narrate the "story" of their product.
  - Create SCCs.
  - List five desired outcomes and then have the client choose the top three.

- *Conduct a contextual inquiry and task analysis for your project.*

  - Immerse yourself in the users' environments and learn how people live and work.
  - Attend meetings or "scrum standups" (discussed in Chapter 3) that let you observe how different stakeholders view the project.
  - Observe what constraints from the work environment impact the tasks users have to perform and how they perform them.

- *Create user stories needed for a "minimum viable product."*

  - Use the "As a < *type of user* >, I want < *some goal* > so that < *some reason* >" template.
  - Examine the stories using the Card, Conversation, and Confirmation technique.
  - Create a story map which shows how the individual stories relate to one another chronologically and in terms of priority.

- *Use think-aloud protocols to observe actual users performing actual tasks.*

  - Use the Informed Consent Statement to create an ethical research study.
  - Avoid language that suggests that participants in the study are being "tested" so they understand that the *product* is being tested, not them.
  - Create scenarios in which participants can identify and choose realistic tasks for them to perform.
  - Screen participants to ensure that they closely represent your target market/user population.

- *Conduct eye-tracking studies of representative users.*

  - Use "heat maps," "gaze plots," and "AOIs" to track users' saccades and fixations.
  - Keep in mind that eyes are faster than tracking technologies, so just because the tracker doesn't show that someone didn't look at something, they may still have seen it.
  - Remember that eye-tracking data tell you *where* someone looked but not *why?* Use think-aloud protocols to understand why users looked at something.

- *Create a SWOT analysis.*

  - Choose participants from diverse departments and teams throughout the organization.
  - Collect data from focus groups, interviews, and questionnaires.
  - Have key stakeholders rank the ideas captured during the research.
  - Map the data on a four-cell grid to show the relationships among ideas.

- *Add more tools to your empathy-building toolbox by learning about other types of research methods to help you empathize with users.*

  - Participate in communities of practice, social networks, and online communities.
  - Attend conferences like UXPA, STC, HCI International, and others.
  - Review online resources such as Usability.gov, the Usability Body of Knowledge, the archives of the NN/g Nielsen Norman group, and popular UX blog sites.

Discussion Questions

1 Upon reviewing the six types of studies highlighted in this chapter, share with a friend or colleague which of these studies you would use to discover specific types of traits and information about your users. For example, which of these is best for understanding users' mental models for completing tasks? Which will help you understand the users' work environments? Which will help you understand users' priorities?

2 How will the data you collect from these studies shape and influence the types of content you create for your users?

3 We have introduced six extremely popular methods of developing empathy for your users, but what are some other methods you might also use? How might you go about finding additional tools for your empathy-building toolbox?

Learning Activity

For this activity, you will need a package of 3"×5" note cards and at least one pad of sticky notes (though it's better if you have three or four pads in different colors). You can do this alone, but it's better to work in teams of two to four people.

Basically, you're going to create a series of user story cards using the template "As a < *type of user* >, I want < *some goal* > so that < *some reason* >." Then you'll build a story map similar to those shown in Figures 4.1 and 4.2.

For this activity, you need to imagine that you're a member of a team that has been tasked with building an online website where people can order and pay for pizzas they pick up at a drive-in window. Begin by creating your note cards. On the front of the note cards, write the user type, goal, and rationale using the template. For example, one of your user stories might be: As a college student, I want to order a bacon, sausage, and onion pizza with extra cheese, so that I can customize the pizza with all my favorite toppings. Next, on the back of the note card, write down all the ways that you can judge whether your story has been successfully completed—i.e., list your "SCCs." For example, one item you might list is that the website needs to provide a menu of extra toppings; it should be provided before your order is complete, and it should include pricing information.

Next, once you've created a set of cards which you think is needed for a "minimum viable product" (in this case, an online website for ordering pizzas), then transfer the names of your note cards to sticky notes. Organize your sticky notes on a whiteboard or corkboard. Use columns to organize the chronological flow of tasks (e.g., customizing the pizza should come before paying for it). Then organize the stick notes into rows within the columns by priority. For example, which is more important—customizing the toppings on the pizza or scheduling a time to pick up your pizza? Again, see Figures 4.1 and 4.2 for examples. Keep in mind that your goal is to begin by building the "minimum viable product" first, so functionality, which is a nice feature but not essential to the overall task, should be listed as a lower priority. You may decide, for example, that certain types of users are less important than others—e.g., you may decide that users who want organic veggies on their pizzas don't need to be supported in the first version of your software.

References

Angeli, E. (2018). *Rhetorical work in emergency medical services: Communicating in the unpredictable workplace.* Routledge.

Beyer, H., and Holtzenblatt, K. (1997). *Contextual design: Defining customer-centered systems.* Morgan-Kaufmann.

Contextual Inquiry (2010). Usability body of knowledge. https://www.usabilitybok.org/contextual-inquiry

Jeffries, R. (2001). *Essential XP: Card, conversation, confirmation.* RonJeffries.com. https://ronjeffries.com/xprog/articles/expcardconversationconfirmation/

Nielsen, J. (2000). *Why you only need to test with 5 users.* NN/g Nielsen Norman Group. https://www.nngroup.com/articles/why-you-only-need-to-test-with-5-users/

Parsons, N. (2021). *What is a SWOT analysis and how to do it right (with examples).* LivePlan.com. https://www.liveplan.com/blog/what-is-a-swot-analysis-and-how-to-do-it-right-with-examples/

Patton, J. (2014). *User story mapping: Discover the whole Story. Build the right product.* O'Reilly.

Ross, J. (2012). *Why are contextual inquiries so difficult?* UX Matters. https://www.uxmatters.com/mt/archives/2012/06/why-are-contextual-inquiries-so-difficult.php

Yarbus, A.L. (1967). *Eye movements and vision.* Plenum Press.

5 Defining Problems and Opportunities

Chapter Overview

In this chapter, we present methods to define user needs based on insights collected from actual empirical research studies conducted by UX writers and designers on development teams. Expanding on the techniques covered in the previous chapter, the specific research methods described in this chapter include (1) persona design, (2) journey mapping, and (3) content inventory/audit studies.

Learning Objectives

- Understand the importance of understanding readers and users in UX writing.
- Create personas that aren't "fictional" but are instead based on hard, empirical facts.
- Develop the ability to create "metrics" for real clients and make informed rhetorical decisions about them.
- Create UX journey maps based on personas.
- Conduct content inventory and content audit studies.
- Create visualizations of data outcomes.

Defining the Challenge

There's a saying that goes, you are already halfway there if you know what the problem truly is. No matter how trivial or complicated (dare we say "wicked") a content design problem is, you need to define its scope and goals clearly. To do so, we recommend three key methods for empirical user research in this chapter, starting with personas.

Real World Snapshot 5.1: Vignettes of UX Writers' Work Environments: Maggie's Job (Continued)

We want to continue with Maggie's scenario that we began in the previous chapter. Maggie, as you may recall, is a UX designer who works for a large, financial services organization that primarily provides processing services for credit card companies, banks, and lending institutions.

DOI: 10.4324/9781003274140-7

Her company has sites all over the country, and Maggie's company has several software and application development teams constantly working both to create new software packages for different types of emerging financial services as well as updating old software to work securely in cloud storage environments, to support new operating systems, or to update network security protocols, just to name a few types of upgrades. At any given time, Maggie's company can have 15–20 major software and application projects underway, and projects will be supported by different types of cross-functional teams. As a result, Maggie's company is divided into departments specializing in the areas of expertise needed. Each member of a team reports directly to the supervisor of their department, and Maggie works in the UX Research department; her job title is "UX designer."

Even though Maggie's title is UX Designer as we saw in the previous chapter, her role begins with conducting research on the end users of a product being created or updated for a project. As the UX Designer, we saw in the previous chapter how Maggie is nearly always responsible for conducting interviews and site visits with end users, conducting think-aloud protocols with users where she observes them conducting tasks with the product and saying what they're thinking as they perform the tasks, or conducting eye-tracking and SWOT analysis studies.

Using the data collected in those studies, we're going to examine in this chapter how Maggie and other UX writers and designers create personas of different user types from the demographic data collected. Then we'll examine the ways that Maggie and professionals in similar roles to hers build on personas to create user experience maps that allow the whole project team to visualize each "touchpoint" where users interact with a product so that the team can design for those experiences, and much, much more.

What Are Personas?

A persona is a composite, archetypal image of different types of users or audiences for your product. A persona is a way of helping authors, designers, engineers, and other members of your design team understand complex demographic information, statistical data, survey results, and other data you have on your audience and users. Personas recognize that instead of saying, "Sixty percent of our users are White females, aged 25–35, and college-educated," it's easier for members of a design team to meet a "realistic archetypal person" who fits the mold. For example, television news stations will create life-size cutouts and photos of people who look like the kind of people who watch their television news broadcasts. They stand these life-sized images by the cameras so that when the newscasters are talking to their audience, they can imagine that these are the people whom they're addressing. It helps them adapt the information they're trying to convey to that type of person. Another example of this happened during the COVID-19 pandemic, when sports stadiums had to be closed to fans due to social distancing guidelines. Instead of playing basketball or football games to empty seats, which made players feel weird, fans could purchase life-sized cardboard versions of themselves to remind players that they still had a real audience and a fan base who were rooting for them.

In the world of UX writing, personas are a way of recognizing that when writers and designers are attempting to address a particular audience demographic, it really helps to "put a face" on the group so that you can truly empathize with them. One of the most famous examples of this is the "Don't Mess with Texas" marketing campaign by the Texas Department of Transportation (TXDOT). In 1985, the TXDOT asked marketing professionals Mike Blair and Tim McClure to create a slogan for an anti-littering campaign. At the time, the state of Texas spent about $20 million annually to clean litter from highways. Research showed that 18- to 24-year-old males were responsible for the bulk of the littering, and so McClure said that "Bubbas in pickup trucks" who regularly littered beer cans and other items out of vehicle windows and ordinary Texans who believed that littering was a "God-given right" were targets of the advertising campaign (Grinberg, 2011, n.p.). The pickup truck-driving Bubba persona "put a face" on the audience the campaign was attempting to reach and helped McClure and eventually his clients understand why slogans like "Keep America Beautiful" or "Give a Hoot, Don't Pollute" simply didn't work with young, male Texans. It helped McClure understand the cultural bias he had to accommodate. He famously said that when his team presented the slogan to their client:

> "The crowd was sprinkled with 'Keep America Beautiful' and 'Keep Texas Beautiful' folks, and our audience is 18-to-24 young males," McClure said. "The 'Keep Texas Beautiful' lady said, 'Can we at least say please?' I said, 'No ma'am, you cannot use the line if you put 'please' in front of it.'"

> (Grinberg, 2011, n.p.)

Personas recognize that when writers and designers are attempting to target an audience, they often base their understanding of the audience on people they actually know. Imagine, for example, that you were asked to write a set of instructions that explained how to share a cell phone photo taken on an Android phone with another cell phone user using text messaging. Now imagine that you were told that you needed to address your instructions to users who were over 60 years old. Where would your sense of who those people are come from? Chances are that your mental model of those users would be based on people you know who are over 60. Grandparents, older colleagues at work, elderly people at your church—the possibilities entirely depend on you. But the point is that your mental image of your audience is probably based on real people whom you've met.

Personas are a means of taking advantage of writers' propensity to base their understanding of what a user or audience needs on a picture of a real person with whom they can identify. A persona not only puts names and faces on users, but it also provides bios, backstories, and sketches about the persona that give an author a nuanced understanding of the users that raw data from demographic information, statistics, and survey results simply cannot. In other words, a persona is a way to tell the story of your users in a more natural and accessible way than a traditional empirical research report or PowerPoint presentation.

In the business world, personas are also an important corrective to an author's individual limitations. As we said previously, if authors are tasked with writing cell phone instructions for people over 60 and if their mental models are based on people they know, then the range of audiences they're likely to address is going to be limited by the authors' own social, cultural, racial, and economic experiences. An author who is, say, an Asian male and who was raised in a traditional, middle-class Asian family from San Francisco will likely have a very different mental model than, say, a White female raised in upper middle-class midwestern suburbia from Kansas City. Although every individual's experience is unique, it's entirely possible that neither of these authors may have had actual experiences with Latinx, African Americans, or other racial and socio-economic groups that might make up a very important demographic within a company's targeted user profile for a product. Personas provide a means of ensuring that the limitations of an author's direct personal experiences don't bias the work toward any single user group at the expense of another. As long as the empirical research studies upon which your personas were based actually target an accurate representative sample of your user population, personas can help authors and companies avoid unintentional bias and exclusionary practices.

So just what does a persona look like? Figure 5.1 below is a sample of a persona from the Usability.gov website developed by the U.S. Department of Agriculture's (USDA) Economic Research Service (ERS).

Persona: Photo:	USDA Senior Manager Gatekeeper

Fictional Name:	Matthew Johnson
Job title/major responsibilities:	Program Staff Director, USDA
Demographics:	51 years old Married Father of three children Grandfather of one child Has a Ph.D. in Agricultural Economics.
Goals and tasks:	He is focused and goal-oriented in a strong leadership role. One of his concerns is maintaining quality across all outputs of programs. Spends his work time: Requesting and reviewing research reports; Preparing memos and briefs for agency heads; and Supervising staff efforts in food safety and inspection.
Environment:	He is comfortable using a computer and refers to himself as an intermediate Internet user. He is connected via a T1 connection at work and dial-up at home. He uses email extensively and uses the web for about 1.5 hours during his work day.
Quote:	"Can you get me that staff analysis by Tuesday?"

Figure 5.1 Sample persona developed by the USDA Economic Research Service.

Source: Usability.gov (https://www.usability.gov/how-to-and-tools/methods/personas.html).

How Do You Use Personas Correctly?

If you're creating websites, software, videos, user manuals, or pretty much any type of media as a UX writer, it should be obvious by now that it is essential to know and understand intimately the needs of your audience. How old are they? How much do they already know about your product or service? What are their goals for using the product or service? What kinds of tasks are they likely to perform? What kind of environment are they in when they're using your UX writing? Are they in a noisy environment where audio is a problem, or can they hear your product? And these are just a few of the types of questions your research will need to answer before you can build a profile of your users. Personas are a way of taking all the data that you've collected to try and address those questions and then compiling that information in a useful, usable way.

One common and very dangerous misconception about personas is that they are "fictional people" or "made-up" characters. This is not the case; *personas are archetypes*, not fictional characters, and it can have severe consequences if members of your design team think about personas as fictional characters. A study conducted by Erin Friess (2012) looked at the ways that a professional design team used personas to make design decisions about a product they were creating. Two people in the design team were actually involved in creating the personas, and they had collected all the data and created the personas the team was supposed to use. But when Friess looked at how often the personas were used to actually make design decisions and who brought up the personas when those decisions were made, she found that they were only used 3% of the time, and they were only mentioned by the two people who had created the personas in the first place (Friess, 2012). The other members of the design team treated the personas like fiction and chose to make decisions based on their own assumptions about the users. They ignored the data and put the project at risk because they could no longer ensure that they were actually meeting the needs of their users by providing positive user experiences. So, when you're creating your personas, it's essential that you involve the other members of the design team in the creation of the personas and that you assure them that the data upon which the personas are based is real.

Another serious no-no a few companies have when they use personas is to create a "library," or a repository, of personas from which designers can choose. They will create prefabricated personas for various character types (e.g., college-educated white female, trade school-educated 63-year-old male union worker, or some other demographic) or for different user roles (e.g., first-time, novice or experienced, legacy user, and so on). The idea is that writers and designers can then pick and choose the personas they want from the repository without having to invest the time and energy needed to create them. However, this library approach is anathema to the purpose of persona-designed products. The entire idea is that the time and energy invested in creating the personas are to provide the team with a *representative* model of their users and audience members. Personas must *never* be used as an excuse for not doing the empirical research upon which the characters are based.

The following are characteristics of a good persona recommended by the XD user experience team at Adobe:

1 Personas aren't fictional guesses of what a target user thinks. Every aspect of a persona's description should be tied back to real data (observed and researched).
2 Personas reflect real user patterns, not different user roles. Personas aren't a reflection of roles within a system.
3 A persona focuses on the current state (how users interact with a product), not the future (how users will interact with a product).

4 A persona is context-specific (it's focused on the behaviors and goals related to the specific domain of a product).

(Faller, 2019, n.p.)

How Do You Create a Persona?

Probably the easiest way to create personas is to look at an example of how one was created for a real situation, so let's start with a scenario.

The Chair of the Department of English at a major Research 1 university was concerned because the number of majors in the English Department wasn't growing at the same rate as other departments at the university. She was worried that maybe their recruiting materials weren't having the impact that the department wanted them to have on prospective students. As a result, she approached the instructor of a content strategy course about having the students in the class research the problem and offer a content strategy program for the department, and a major part of that strategic plan necessarily involved creating personas for prospective students.

To begin the strategic plan, the students first needed to figure out where majors the in the department came from. How did they find out about schools, what channels do they use to seek out information about majors, how did they ultimately decide to choose the major, and other basic information about the audience we needed to reach? Using many of the same techniques that we discussed in the previous chapter, the students decided to conduct surveys and interviews with alumni from the department, parents of students in the department, current students in the department, high school counselors in the surrounding area, faculty in the department, and administrators in the department. They also conducted content audits of the department's public-facing materials, which we will be discussing later in this chapter.

In the process of collecting all this data about who the majors in the department were and where they came from, the team made a surprising discovery. Most people, including faculty in the department, assumed that new majors entered the department straight out of high school. They assumed that the best way to recruit them was to attempt to reach them through high school counselors, career fairs, direct mail to students who take the Scholastic Assessment Test (SAT), etc. In fact, however, what we discovered was that over 60% of the majors in the department actually switched majors to English during either their freshman or sophomore years.

Through our interviews, we learned that over 60% of the majors in the department chose the degree because they had taken a sophomore literature course or a writing course early in their academic careers and discovered they had an aptitude for it. Often, they reported that their Biology, Civil Engineering, Computer Science, Psychology, or other major wasn't "what they expected it to be," and so they looked for an alternative major. In many cases, current students and alumni told us that they had always wanted to develop the skills demanded of English majors but that their parents discouraged them because they wanted them to become doctors, engineers, or technicians. Neither the prospective students nor their parents realized just how many career paths were available for English majors or that over 60% of graduates of the program had jobs in less than three months. The survey of alumni from the major showed the following breakdown of career paths:

- 6.9% – Volunteer nonprofit
- 6.9% – Advanced graduate studies

- 6.9% – Federal, State, and, Local Government
- 10.3% – Technical Writing
- 13.8% – Professional and Business Services
- 20.7% – Media and Entertainment
- 34.5% – Corporate Training and Educational Services

Our background research needed for creating personas also revealed that, not only was it a myth that English majors could only get jobs as teachers, but it also showed that graduates were very successful at finding jobs before or soon after graduating. Almost 63% of graduates had positions within zero to three months of graduating, and closer to 75% had positions within six months. All of these discoveries were startling revelations, even to senior faculty and administrators in the department.

These discoveries, particularly the discovery that most of the department's students came from other programs at the university, meant that we really needed to create two different personas. One persona that reflected the traditional, "declared major" who joined the department right out of high school. The other persona was needed for the "transfer students," who typically entered the department before their junior year.

Once we had determined how many types of personas we would need, the next problem that we had—and one you almost always have when you're doing the research necessary to put together solid personas—was how to deal with the amount of data collected. In this case, the amount of data we had collected was truly massive. When we put all the information that we collected into a slideshow, it took over two hours to present and had over 97 slides in the presentation. The students in the Content Strategy class were so excited to share *everything* they had learned that they actually made the mistake of overwhelming our client with so much information that the client couldn't decide what was actionable research information and what was merely "nice to know."

Fortunately, personas allow you to get around that problem. They enable you to condense and synthesize the information down to, usually, a single page.

Getting it all on one page, however, means that you must make critical and often difficult decisions about what "categories of information" should have the highest priorities. What, in other words, are the most important things your UX writers need to know, and what information categories have a lower priority? The question here is difficult because the information categories you come up with for each project you do are necessarily going to be different depending on the goals of the project.

For example, if you're doing a documentation project providing instructional videos, tutorials, and/or manuals for a software package, you're probably going to need to emphasize how much the user already knows about the product, whether they are a legacy user or a first-time user, the environmental conditions under which the user is operating, their pleasure/frustration level with the product, and so on. In the case of our work for the English Department, our interest was primarily in the information that would help us reach and recruit prospective students to the program. Consequently, the categories of information we needed to track on our personas differed significantly from those we would have used if we were writing a manual.

As Figure 5.2 shows, we didn't include the information we collected from alumni about what types of jobs they had or how long it took to obtain them. Although this information was useful and we discussed it at length when we were describing the type of content that should be included on flyers, brochures, and websites for the department, it wasn't helpful in terms of the goals for our personas. The goal of the information that needs to appear

Major Transfer

Alison Miller

Bio

Alison started out at Clemson as a Chemical Engineering major because she felt pressure from her parents & older brothers to become a STEM major. Although she always felt more passionate about reading and other creative endeavors, she felt as though she would not be as successful. After realizing thae chemical engineering was not the right choice, Alison decided to follow her passion and changed her major to English.

Demographics

Age
20

Family
Parents are married, 2 older brothers, middle class

Home Town
Livingston, NJ

Goals

- Be passionate about her major
- Take classes that will help her get a job
- Get a job after graduation

Frustrations

- Justifying her major change to family
- Worrying about graduating on time
- Concerned about fewer job opportunities

Motivations

Wants to graduate with a degree in English

Wants to gain valuable skills that directly relate to her career goals

Wants to use the skills she learned to get a job after graduation

Preferred Channels

- University Admissions Website
- English Dept. Website
- External Research

Quote

"Even though I loved English in high school, I felt pressured to pursue a STEM career. But I realized that I didn't have the passion for Chemistry, so I transferred to a major where I could be both happy and successsful."

Influences

- Parents (most important)
- Soph. Lit. Professors
- College Advisors

Figure 5.2 Persona of a transfer major in the English Department.

on a persona is to describe each archetypal image in such a way that writers and designers develop a deep understanding and empathy for the users and/or audience.

According to the website *Usability.gov*, typical elements or categories of information you decide to include in your personas might be as follows:

Elements of a Persona

Personas generally include the following key pieces of information:

- Persona Group (i.e., web manager)
- Fictional name
- Job titles and major responsibilities
- Demographics such as age, education, ethnicity, and family status
- The goals and tasks they are trying to complete using the site
- Their physical, social, and technological environment
- A quote that sums up what matters most to the persona as it relates to your site
- Casual pictures representing that user group

(*Personas*, n.d.)

Personas like the one shown in Figure 5.2 are usually based on one of the many persona templates available on the web. Many of these are available from free or "freemium" sites, and the bulleted list below provides a list of some of the popular sites that students in our courses have used successfully. In fact, the template students used to create the persona in Figure 5.2 was based on one of the several types of templates found at Piktochart listed below.

- Xtensio User Persona Template – https://xtensio.com/user-persona-template/
- Random Face Generator/This Person Does Not Exist – https://this-person-does-not-exist.com/en
- Free Adobe User Templates (login required) – https://www.adobe.com/express/create/user-persona
- JUSTINMIND 30 must-see user persona templates – https://www.justinmind.com/blog/user-persona-templates/
- Piktochart (login required) – https://create.piktochart.com/
- Atlassian (login required) – https://www.atlassian.com/software/confluence/templates/persona
- Venngage's templates (login required) – https://venngage.com/blog/user-persona-examples/
- Hubspot Interactive Make My Persona Tool – https://www.hubspot.com/make-my-persona
- Random Name Generator – https://www.name-generator.org.uk/

What Are Journey Maps?

Journey maps (also known as "user experience maps," "customer experience maps," or simply "UX maps") are like personas because they provide you and your development team with a visual compilation of enormous amounts of data. Jim Kalbach actually covers five different types of what he calls "alignment diagrams" in his book *Mapping Experiences* (2016). An alignment diagram is "a *category* of diagram that illustrates the interaction between people and organizations" (p. 4). He describes maps such as "system blueprints" that teams can use when they want to look at the infrastructure that needs to be in place to support systems and services. If, for example, you worked for a company like FedEx, a service blueprint could be used to let you know that you need a driver, a fully fueled vehicle, and a map to a destination as just some of the infrastructure that needs to be in place before a delivery can occur. In addition to service blueprints, Kalbach also includes customer journey maps, experience maps, mental model diagrams, and spatial maps as types of alignment diagrams a product development team might use (pp. 6–11) to visualize interactions between people and organizations. But for our purposes, we want to focus on traditional journey maps or user experience maps.

What Are Key Elements in a Journey Map?

Basically, a journey map tracks and provides information about every opportunity the user has to interact with your company's product or service. These points of interaction are called "touchpoints." The touchpoints are organized chronologically underneath what is called either the "epic" or sometimes the "backbone." The epic (or backbone) is basically just the phase in which the touchpoints occur. For example, there's nearly always an exploration or discovery phase in a journey map that shows how the persona located the product in the first place. For example, if we were mapping the user experience of ordering and picking up a pizza from a major pizza chain, the persona would likely have had several opportunities to discover the company's services, e.g., television commercials, billboards, radio commercials, websites, Facebook posts, and many others. All these "touchpoints" would be collected under the "discover phase" in the epic.

It's also important to note that journey maps are built on the basis of a single persona and a specific "scenario" that provides a unique "point of view" to the journey map. Below is a case study from USA.gov and an example of a persona (named Linda) and her scenario. In this case, Linda wants to browse information on the USA.gov site so she can learn more about how to find financial assistance from the government. The authors of the case study explained that they selected "searching for financial assistance from the government" as the topic of Linda's journey "because it's consistently one of the top reasons customers visit USA.gov or call 1-844-USA-GOV1" (Monroe & Chronister, 2020, n.p.).

Figure 5.3 illustrates Linda's persona and her scenario.

As we discussed in the previous section of this chapter, just as the persona we discussed in the previous section of this chapter was based on actual empirical data you collected from your interviews, surveys, and other research on your users, the persona and scenario here are based on actual research. The persona and scenario are essential parts of the journey map because they help readers understand the point of view from which the journey map needs to be read. A different persona will provide a different point of view with different touchpoints and experiences. The scenario contributes to our understanding because it provides readers with an understanding of the expectations and goals the persona has for interacting with the website, product, or service. In Linda's case, the scenario makes clear that she's using the USA.gov website in order to find out information about financial aid that can help her because she's lost her job and she no longer has a husband upon whom she might have been able to depend. The scenario, when combined with the persona, makes Linda's point of view clear and helps UX writers and designers empathize with her needs.

Once you have the persona and scenario, it's much easier to understand Linda's journey through the USA.gov websites shown in Figure 5.4 and why her journey ends in

Browse information or learn more on a general topic

Linda's husband passed away two years ago and she's been struggling to make ends meet ever since he died. She was working as a contact center representative, but recently lost her job because her company downsized. She is worried about how she will support herself and is frantically looking for financial assistance until she can get a new job.

A friend of Linda's told her to look for government grants. Linda uses a computer for email and Facebook, but isn't great at finding information online. She did a Google search for government grants and clicked on the first result. Linda is confused about what grants are available to her.

Needs:
- Help finding information online
- Easy to understand information
- Financial support from the government to help pay her bills

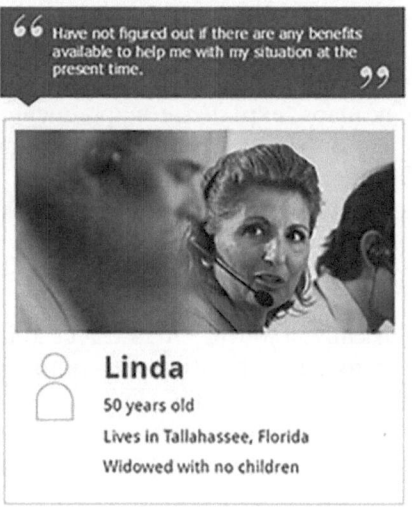

> 66 Have not figured out if there are any benefits available to help me with my situation at the present time. 99

Linda
50 years old
Lives in Tallahassee, Florida
Widowed with no children

Figure 5.3 Sample persona and scenario from USA.gov.

Source: Monroe and Chronister (2020).

disappointment. At the top of Figure 5.4, you can see the reference to the Linda persona and her scenario. The next line down is labeled "Stage of Journey" and represents the epic line we discussed previously. The line below is labeled "Activities," and it is the line we were calling "touchpoints." In this case, you can see how the phases in the epic line can contain multiple related touchpoints. Linda has three different touchpoints, for example, listed under the phase "Looks for Information," and she only has one touchpoint under "Identifies Information Need."

Below the persona + scenario, the epic, and the touchpoint lines are what are known as the "metric" lines. The metrics are where you can become very creative with your journey maps, and they're where you can show information about the journey that your team can use to make critical decisions about the experience you wish to create for your users and audiences. Traditionally, one of the metrics you will probably want to include on your map is something about the feelings or emotions the persona is experiencing at a particular touchpoint. Figure 5.4 does this in two ways: first, it shows frowny and smiley emoticons to provide an iconic representation of Linda's emotional state at each touchpoint. This is combined with a line graph illustrating how happy or sad Linda is at each point. There is also a textual explanation of Linda's feelings and needs that aids in the interpretation of the emoticons and line graph. Journey maps can often have from three to five different metrics that the design team believes are important to understanding each persona's needs and experiences.

Finally, the bottom line of a journey map is traditionally called the "opportunities" or "takeaways" line. As the example in 5.4 shows, it lists opportunities the design team has for creating new products and/or improving experiences on current products based on findings illustrated in the journey map.

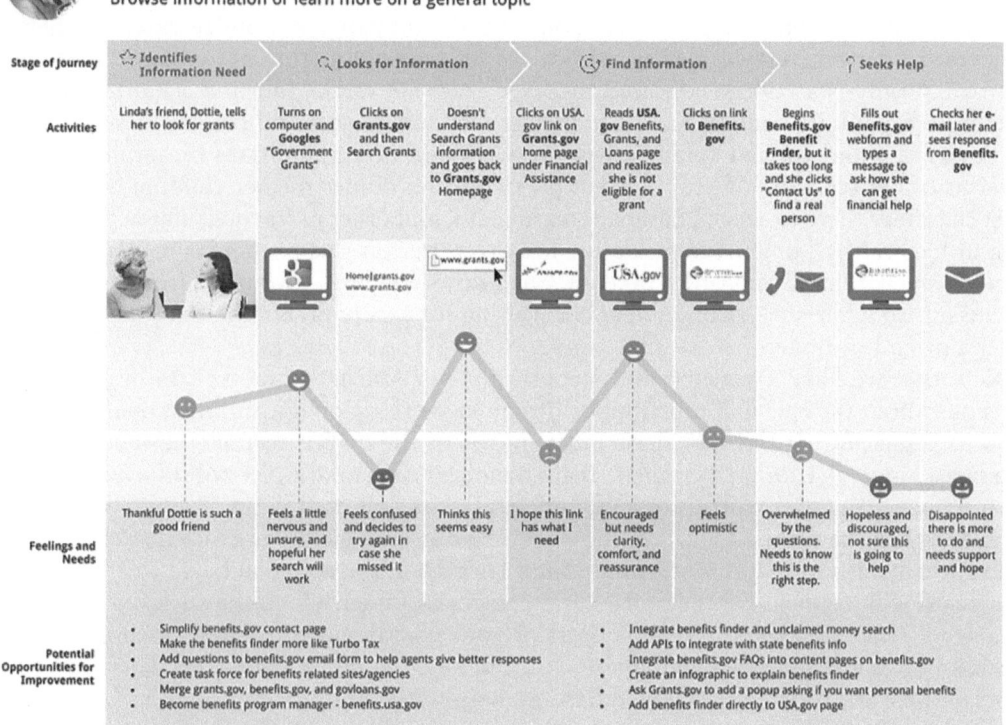

Figure 5.4 Linda's journey map from USA.gov.

Source: Monroe and Chronister (2020).

Collectively, the elements from top to bottom in a journey map are:

1 Scenario + Persona
2 Epic (or backbone)
3 Touchpoints
4 Metrics
5 Opportunities (or takeaways)

How Do You Create a Journey Map?

Many people are intimidated at the thought of creating a journey map because the only examples they've seen are highly refined final products that usually have lots of colors, icons, and imagery on them, like the example above. But the actual process of building a journey map can be fun, messy, and very creative.

The first step in the process is to first select your persona and build your scenario. The reason for this will become immediately apparent if we return to our previous example of ordering a pizza. How are we going to determine the phases our map is going to need if we don't know whether our user/persona needs to order the pizza on the phone, in person at the store, or using an online app? The persona and scenario provide us with the goals we need to address questions like this.

For the sake of our ordering pizza example, let's assume that we've done market research on customers who purchase pizzas from our client's company, and our research has helped us develop a persona named Wendy. Wendy is a 38-year-old mother of two boys, ages seven and ten. Wendy is a college-educated divorcee, and she works full-time as an accountant at a large community college. She's a proficient technology user and is accustomed to using her Android phone to read email, search the web, and order purchases from online stores. In terms of her scenario, Wendy is often pressed to fix dinner for her children because she must chauffeur them to soccer, baseball, math team, and other extracurricular activities. She tries to fix healthy homemade meals for her boys, but on this particular day, she decides to give herself a break and order takeout. She knows her boys love pizza from a particular restaurant, so while she's sitting in her car waiting for soccer practice to end, she decides to order a pizza for dinner.

Now that we have a persona and scenario for our journey map, we can begin making decisions about the epic and touchpoints. If you're working on a team, then one of the best ways to begin putting together these initial pieces of the map is to use color-coded Post-It notes (see Figure 4.1 for an example). Team members each take a piece of information from the interviews, questionnaires, surveys, and other data collected, write it down on a Post-it note, and then the team attempts to locate it on a chronologically organized map.

Eventually, as more and more data points are added, the epic will begin to emerge, and your team will begin to see the phases and categories under which the sticky notes can be organized. Once you've got a rough draft of your epic mapped out and a sense of which touchpoints might go under each phase, you can begin building your map. There are commercial software tools you can purchase to help with this process, such as Canvanizer, Touchpoint Dashboard, and others. However, our experience shows that a simple spread-sheet tool like MS Excel or Google Sheets does an admirable job of laying out the bare-bones outline of your map. Figure 5.5 below illustrates how Wendy's experience ordering a pizza might look.

Figure 5.5 also illustrates how you can begin making decisions about what types of metrics to include on your journey map. For example, in Wendy's pizza example above, we can see that, during their interviews with real customers, the team had begun collecting information about Wendy's frustration levels at different points in the process. The users complained about not being able to download the application directly from the store's main website and being forced to go to the Google Play store to obtain the application. They also complained about being forced to create an account on the app and having to provide personal information such as their credit card and physical location before they could use the app. All this data suggests that at least one of the metrics in the map of Wendy's pizza ordering journey could be the same use of emoticons to indicate her feelings as we saw in Linda's journey map in Figure 5.4.

For large commercial clients, however, smiley faces and emoticons can tend to trivial-ize the data and appear "childish" to the client. Consequently, you may want to consider collecting other, more quantitative measures for your metrics. Later, in Chapter 8, we will be discussing measures you can use to assess the success (or failure) of your products for users, but the bulleted list below provides three commonly used metrics large commercial organizations use to measure customer experiences.

- **Net Promoter Score (NPS)** – how likely a customer is to recommend a brand to others

persona + scenario	Wendy (persona) wants to order a pizza online, pick it up at the store, and bring it home (scenario)																	
Epic (aka "backbone")	Discovery						Placing Order					Completing Order				Receive Pizza		
touch-points	Sees tv adverts	Decides to order on android phone	Goes to main website	Downloads app & installs on phone	Creates required account	Selects nearby store	Review menu	Review special offers	Chooses pizza style	Chooses toppings	Adds to cart	Selects pick up time	Selects payment type	Enters credit card	Chooses pickup over delivery	Arrives at drive up window	Provides last name to cashier	Recieves pizza + receipt
Metric 1 (e.g., level of frustration)			satisfied she found site	annoyed she has to go to Play Store to locate app	dislikes so many steps to get started			likes prices								annoyed there's no ringer-buzzer to get help		happy she's done, but wishes she'd asked for napkins
Metric 2																		
Metric 3																		
Opportunities																		

Figure 5.5 Ordering a pizza organized using spreadsheet software.

- **Customer Effort Score (CES)** – how much effort customers have to exert to reach their goals
- **Customer Satisfaction Score (CSAT)** – usually a basic Likert scale running from extremely dissatisfied to extremely satisfied

(adapted from Clinehens, 2022, n.p.)

Closing the Loop on the Journey Map (for Our Client)

Now that we've discussed the purposes for creating journey maps, key elements included in traditional journey maps, and techniques for creating journey maps, it's time to return to the scenario with which we began this discussion. As you'll recall from the example we used to create our persona, our client was the Chair of the English Department, who was concerned because the number of majors in the English Department wasn't growing at the same rate as other departments at the university. She asked us to devise a content strategy plan that would provide effective recruiting materials for prospective students. After research, the persona we developed as the archetype for students transferring from other majors was named Alison. As Figure 5.2 showed, Alison started out as a chemical engineering major due to pressure from her parents and siblings, even though she was more passionate about the creative work she'd done. After her first year of coursework, she realized that she needed to change majors. In the scenario for Alison's journey map shown in Figure 5.6, Alison's goal is to find a major in which she can both pursue her passions and be successful.

What Are Content Audits?

As the term "audit" suggests, a content audit is a listing of all the digital content that an organization provides. Content audits usually consist of two parts: the first is the "content inventory," which lists all the channels the organization uses (e.g., website, Twitter, blogsite(s), Instagram posts, and Facebook pages). The second part is a detailed inventory of the specific content on each of the channels. Content inventories track what kinds of content exist in a system. Content audits assess the quality of the content.

As the Usability.gov website explains, content inventories vary in what they capture, but most include:

- Unique Content ID
- Title
- URL
- File Format (HTML, PDF, DOC, TXT...)
- Author or Provider
- Physical location (in the content management system, on the server, etc.)
- Meta Description
- Meta Keywords
- Categories/ Tags
- Dates (created, revised, and accessed)

(*Content Inventory*, n.d.)

In the case of the English Department's search for a content strategy that would improve their recruiting, the students working on the project created an inventory of the department's website using an Excel spreadsheet, as shown in Figure 5.7.

Internal Transfer Journey Map

Phase	Stage	Expectations	Thinking	Feeling	Opportunity for Improvement
Consideration	Discusses school choice(s) w/ family	High-school counselors to know this information	I wonder if these schools will have what I'm looking for	Nervous about a new school.	Brochure's and Tri-Fold Flyers sent to high-school counselors.
Consideration	Applies to Clemson under an English major	This to be the right major	I hope I can get in under this major, it is so competitive	Hopeful and excited.	Give information about English major to all Clemson advisors • Add statistics to Clemson English website
Consideration	Attends Clemson	College to be as easy as high-school	This is a lot harder than I thought	Happy to be at such an inclusive school.	
Consideration	Struggles with classes under first major	To be able to get grades up by end of semester	I don't think this is the major for me	Worried I picked the wrong major	
Consideration	Talks with parents about change in major	Parents to know a lot about English as a major	I hope my parents can help me with this big decision	Hopeful parents can help me	
Consideration	Parents are unsure, tells them to speak to advisor	Parents to help pick a direction	Why can't my parents understand that I can get a job from an English degree too	Worried I won't be able to become an English major	
Consideration	Makes appointment with Keri who relays info.	Current major advisor to know something about English dept.	Maybe my advisor can help me	Nervous because my advisor didn't know what to tell me	
Consideration	Emails Keri about change in major	Keri to be very helpful	Hopefully this Keri person will get back to me quickly	Nervous that English isn't what I'm looking for in a career.	
Research	Attends meeting with Keri	Keri to get me all settled in new major during meeting	This information is what I've been looking for	Getting hopeful I can convince my parents	Maintain strength of communication with Undergraduate Advisor • Regularly update website and social media content
Research	Speaks with parents again and relays information	Parents to feel better about English as a major	I'm a little apprehensive that this is the right major for me	Excited parents reacted well to new info Keri gave me	
Research	Decides to try to switch to English major	This will be a quick process	I hope I made the right choice	Hopeful that this new major will be smooth.	
Decision	Fills out 'change in major' form on degree works	This to be the final step	This process was a lot more difficult than it should have been	Happy to be done with the process	
Decision	Submits form and waits on approval	This to be the final step	Why couldn't Keri approve this	Glad to finally start the English major	
Confirmation	Gets approval	Approval to be given with Keri in steps before	I hope this major goes better for me	Excited for a new start	Add all important information about courses to social media outlets • Have clear link to Course Description on Clemson English website
Confirmation	Register for Classes	To get all the classes I need/want	I wish this information for course descriptions were close to this link	Ready to give this major a go	

Feeling axis: Positive / Negative

Figure 5.6 Map of Alison's decision to transfer majors.

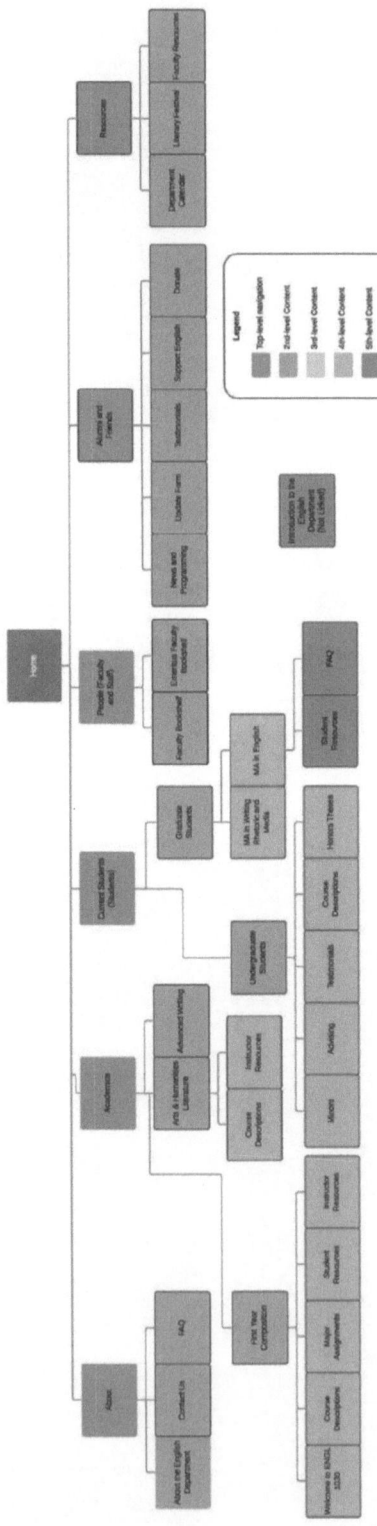

Figure 5.7 Excerpt from an Excel spreadsheet showing the inventory of the English Department. website.

After the inventory, the second part of a content audit involves assessing the quality of the material in the inventory. Often called the "qualitative audit," this part of the audit is used to "uncover content that needs updating, [to show] where gaps exist that new content could fill, and [to determine] if certain pieces of content are ready for removal" (Kaley, 2020, n.p.). For example, using the content inventory of the English Department website in Figure 5.7, students were able to produce a wireframe map of the site that showed pages that were buried five layers deep on the site and, worse, "orphaned" pages that lacked any internal links to them, rendering them undiscoverable to users (Figure 5.8).

According to the Nielsen/Norman Group, some additional factors that should be considered during the qualitative audit include:

- **User needs:** Specify the audience, its task, and its needs. Who are your content users, and what are they trying to do (e.g., find answers, discover new information, learn about new topics, compare options, make a decision, and get in contact)? To what degree does the content support them in that task? Do they have any unanswered questions?
- **Content standards:** To what degree does the content reflect the organization's intended tonal values, include appropriate metadata, follow formatting and structuring guidelines, and uphold design principles?
- **Goals and performance metrics:** State what the content is supposed to be doing (e.g., create awareness, drive traffic, generate leads, and sell something). Use performance metrics, such as clicks, views, bounce rates, likes, and shares, coupled with any qualitative insights from user research, in your analysis. Does the content help reach the goal or detract from it?

(Kaley, 2020, n.p.)

The N/N Group recommends assessing these factors using a rating system that gives a "High," "Mediocre," or "Low" score for the level of content quality given best practices, internal content standards, user needs, business goals, and metrics (Kaley, 2020).

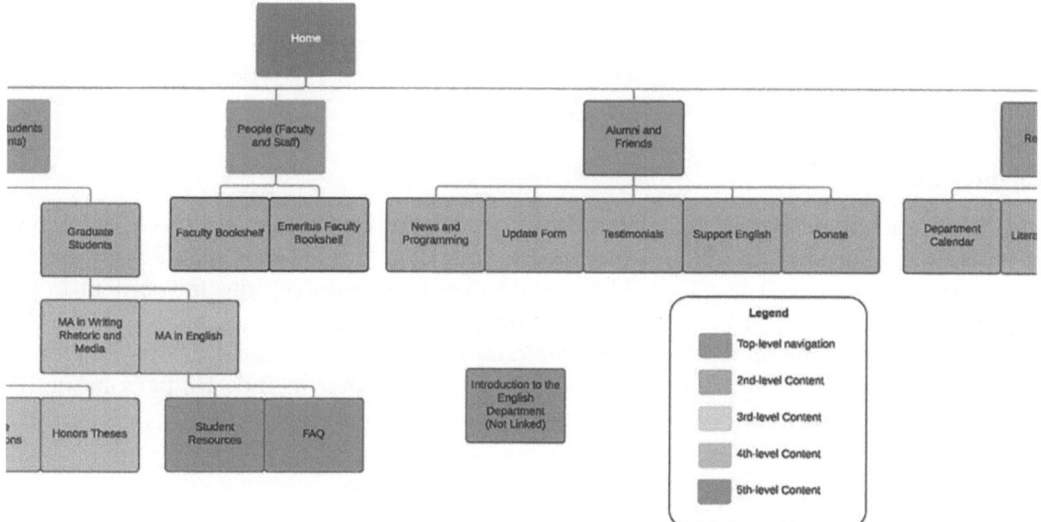

Figure 5.8 Excerpt of a wireframe showing orphaned and buried pages on a website.

Fortunately, there are many software and web applications available to assist you with creating content inventories and qualitative audits. The following are just a few software packages that are available:

- Google Analytics – https://marketingplatform.google.com/about/analytics/
- SiteAnalyzer – https://www.site-analyzer.com/
- Website Grader (from Hubspot) – https://website.grader.com/
- DynoMapper – https://dynomapper.com/
- Screaming Frog – https://www.screamingfrog.co.uk/seo-spider/
- Web Gnomes – http://www.webgnomes.org/

Often, it's useful to use a combination of these tools to conduct your qualitative content audits. Indeed, the students used a combination of Google Analytics and Screaming Frog to create the spreadsheet shown in Figure 5.7.

Conclusion

This chapter piggybacked on the research methods introduced in Chapter 4. It assumes that, when you're working on real projects, you will have already used at least some of the research methods we presented there. To create personas and UX maps, you must begin by defining your users' needs based on insights collected from those empirical research studies. In this chapter, we then introduced the three most popular methods for building a well-defined content strategy plan by visualizing data collected about users. These methods were (1) persona design, (2) journey mapping, and (3) content inventory/audit studies. These methods are so popular in the workplace that, when you hear other professionals talking about "persona design" and journey mapping, they often fail to make clear that they're collecting data as part of their processes (or they're making up stories based on their personal experiences that they shouldn't do). Either way, it's critical to remember that all three methods (and most particularly personas and journey maps) are based on *real* data collected from actual users. As long as they're performed correctly, personas, UX maps, and content inventories/audits are extremely powerful tools that will enable you and your UX writing team with what you need to make informed decisions about the types of content and the best ways to present it in your content strategy plan.

Chapter Checklist

In UX writing and designing, it's often necessary to go beyond merely collecting data about your users. You need to "package" and present your data in ways that enable both you and your team to define your users' problems and to see new opportunities for content in particular ways. This chapter has focused on three popular techniques for analyzing the data you collect. Expanding on the specific research methods covered in the previous chapter, the analysis and visualization techniques described in this chapter include (1) persona design, (2) journey mapping, and (3) content inventory/audit studies.

- *Use research from studies in Chapter 4 to create personas.*
  - Base your personas on *actual facts* rather than fiction.
  - Create personas for *all* the members of your team.
  - Use metrics for your personas that address specific choices your engineers, designers, and graphic artists need to make.

- Avoid cartoon images and cutesy names that trivialize your personas and make them less believable.

- *Create journey maps for each individual persona you create.*

  - Create a "backbone" or "epic" that breaks the user's journey into beginning, middle, and end phases.
  - Find all the "touchpoints" users have with the product and map them under phases in your epic.
  - Develop metrics to use under each touchpoint that will help you and your design team understand what users are experiencing, thinking, and feeling at each touchpoint.
  - Use Excel or some other spreadsheet software to create a mockup of your journey map before going to the expense of adding colors and graphics.
  - Include a line at the bottom that illustrates how the map shows opportunities to revise the product, add new features to a product, or even introduce entirely new products.

- *Conduct a content inventory.*

  - Use tools like Google Analytics and Screaming Frog to create an inventory of all the content.
  - Use a spreadsheet to describe and track the inventory.
  - Create metrics for the spreadsheet that track information such as when the content was created, when it was last reviewed, when it was last revised/updated, how long it takes to load, etc.

- *Create a content audit.*

  - Use the content inventory to examine each piece of content.
  - Evaluate the content in terms of meeting audience needs, the organization's look-and-feel guidelines, and relevance to the product.
  - Use performance metrics to score quantitative measures like the number of clicks/views, bounce rates, likes and shares, NPS, CES, and/or CSAT.
  - Score qualitative content using a rating system that gives a "High," "Mediocre," or "Low" score for internal content standards.

- *Add more tools to your content auditing toolbox by learning about other types of metrics you can use to measure content quality.*

  - Participate in communities of practice, social networks, and online communities.
  - Attend conferences like CHI, UXPA, PRSA, HCI International, and others.
  - Attend workshops offered by organizations like the STC, the NN/g Nielsen Norman group, the UXPA, and others.

Discussion Questions

1 Upon reviewing the three types of studies highlighted in this chapter, share with a friend or colleague which of these studies you would use to address the following questions:

   - Which is best for understanding users' mental models for completing tasks?
   - Which will help you understand the users' work environments?
   - Which will help you understand users' priorities?

2 How will the information you display in personas, journey maps, and content audits shape and influence the types of content you will create for your users?

3 Consider the journey map shown in Figure 5.6. Discuss with friends or colleagues how a UX writer/designer would use the information shown at each touchpoint to generate content for a website or some other information product (e.g., a social media entry, a flyer, a brochure, a digital video, etc.).
4 We have introduced three extremely popular methods for developing empathy for your users, but there are new methods being developed all the time. How might you go about finding additional tools for your UX writer's toolbox?

Learning Activity

You may do this activity on your own or in teams of two or three. Begin by selecting a website that you know well and use regularly. Beginning with the homepage for the site, create an Excel spreadsheet or some other spreadsheet that will provide an "inventory" of content on the site (see Figure 5.7 for an example). For your inventory, track the following pieces of information on each page:

- URL path
- Page title
- Description of the page
- Media types on the page
- Word count
- Links to other internal pages
- Links to external sites

Consider the audience/users of the site. For each page, evaluate the content in terms of how well it meets the audience's needs. Score how well it meets the needs using the "High," "Mediocre," or "Low" rating system. Record your score on the spreadsheet. Next, on the spreadsheet, provide a rationale that justifies your score.

Next, consider the content's relevance to the product for each page. Again, score the content on the "High," "Mediocre," or "Low" scale. On the spreadsheet, provide a rationale that justifies your score.

Consider other criteria you might use to assess the quality of the content on each page (e.g., timeliness and up-to-date information). Continue adding this information to your spreadsheet.

References

Clinehens, J. (2022). *Customer experience measurement: How to use metrics in your journey map.* CX That Sings. https://cxthatsings.com/journey-mapping/customer-experience-measurement/

Content Inventory (nd). Usability.gov. https://www.usability.gov/how-to-and-tools/methods/content-inventory.html

Faller, P. (2019, Dec 17). *Putting personas to pork in UX design: What they are and why they're important.* Adobe XD. https://xd.adobe.com/ideas/process/user-research/putting-personas-to-work-in-ux-design/

Friess, E. (2012). "Personas and decision making in the design process: An ethnographic case study." ACM CHI '12. *Archival Research Papers on Human Factors in Computing Systems*, pp. 1209–1218.

Grinberg, E. (2011, July 1). *Why there's no messing with Texas.* CNN. https://web.archive.org/web/20110714005926/http:/articles.cnn.com/2011-07-01/us/texas.pride_1_texans-bumper-stickers-texas-department?_s=PM%3AUS

Kalbach, J. (2016). *Mapping experiences: A complete guide to creating value through journeys, blueprints, and diagrams.* O'Reilly Media.

Kaley, A. (2020, Sept. 27). Content inventory and auditing 101. NN/g Nielsen Norman Group. https://www.nngroup.com/articles/content-audits/

Monroe, M.A., and M. Chronister (2020, Sept. 8). *Journey mapping the customer experience: A USA.gov case study.* Digital.gov. https://digital.gov/2015/08/12/journey-mapping-the-customer-experience-a-usa-gov-case-study/

Personas (n.d.). Usability.gov. https://www.usability.gov/how-to-and-tools/methods/personas.html

6 Ideating and Prototyping Content

Chapter Overview

This chapter teaches you how to generate radical content solutions by using UX methods—i.e., card sorting, affinity diagramming, participatory design, and the 6:1 and four-category methods. Then, we discuss several prototyping strategies, including hi-fi and lo-fi prototyping, to materialize the ideated solutions for testing later.

Learning Objectives

- Understand the ideation process to generate strategic and radical content solutions.
- Conduct card sorting, affinity diagramming, and participatory design exercises.
- Generate lo-fi to hi-fi content prototypes.
- Create paper prototypes, wireframes, and mockups for testing.

What to Do with Defined Problems?

In Chapter 4, you were introduced to the ways writers and designers seek information about users through methods like interviews and contextual inquiry. In Chapter 5, we provided some strategies for scoping down a specific focus for the design project. Through personas, journey mapping, and content auditing, you can start to clarify the direction for design. We called this process *problem definition*.

Problem definition is also the bridge between research and design in UX writing. Refer to the "double diamond" model in Figure 6.1. The left diamond represents the research phase—where practitioners first *diverge* in discovering the potential design problem and then *converge* to define the scope of the problem—and the right diamond represents the design phase. You will notice that problem definition joins the two and leads to a second divergent thinking exercise where writers *think big* to generate innovative ideas that can later be turned into actual solutions. This phase is called *ideation*.

Using the problems writers have defined in the previous phases of research, ideation consists of multiple activities to develop ideas that later get made into actual, testable solutions through the prototyping exercise. Let's begin with a key mentality for the ideation process—divergent thinking.

DOI: 10.4324/9781003274414-8

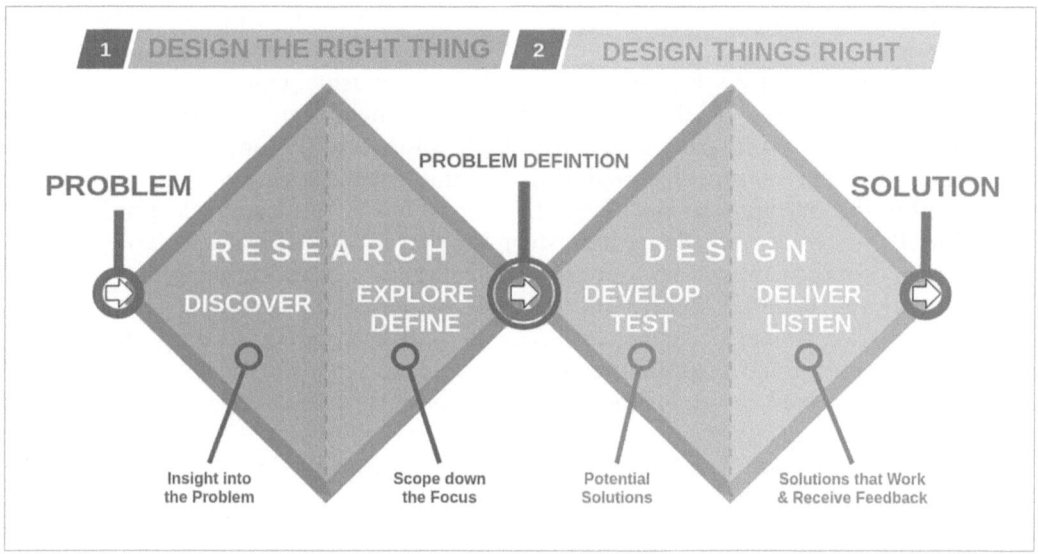

Figure 6.1 The double-diamond model for UX design. The model was popularized by the British Design Council in 2005.

Source: Wikimedia.

Ideation: A Divergent Brainstorming Process

Ideation is an exciting process. Many writers look forward to this part of their design project. But first, you should keep in mind that the goal of ideation is to generate a large number of ideas—high-quality, low-quality, and even mediocre ideas. So, it is important to practice divergent thinking. Mentally, you want to branch out into *many ideas*, *different ideas*, which can lead to *better ideas*. However, writers, especially when working in teams, may fall prey to a social phenomenon that communication scholars call "groupthink" (Janis 1991). Groupthink happens when a team unanimously agrees to one idea that someone from the team presented without considering other possibilities first. Roger von Oech, in his book, *The Creative Contrarian* (2021), curated a host of "wise fool" strategies to combat groupthink for the reason that groupthink can lead to poor decision-making— like the well-known *Challenger* space shuttle disaster of 1986, explained in Irving Janis's work. Essentially, NASA decision-makers and top echelons were more concerned about meeting the expectations of its stakeholders and the American public than the safety of the launch and its crew members. This pressure came from national and political forces—even then-President Ronald Reagan had announced the launch at his Union address. Rather than listening to space shuttle engineers on their calculations, NASA and solid rocket booster manufacturer Morton Thiokol felt compelled to go ahead with the launch as scheduled, a faulty decision that led to seven astronauts losing their lives when *Challenger* broke apart 73 seconds into its flight. In the case of UX writing, groupthink can result in uncreative or inappropriate content solutions. Practicing divergent thinking can help avoid groupthink.

Ideation provides both the fuel and the source materials for building prototypes later. At this stage, you shouldn't be evaluating the quality or viability of ideas but instead focus on going wide and going wild with your imagination. Remember not to settle for early and easy solutions. The goal of UX writing is to innovate content that best serves user needs and client goals. It's inevitable for you to arrive at obvious solutions during the early part of ideation, so it is OK to write them down and set them aside. Once you have gotten the obvious ideas out of your head, you can focus on the unconventional ones. For the same reasons you want to avoid groupthink, you don't want to settle for early and easy ideas. You want to avoid premature closure during this brainstorming phase.

It is within the attributes of the design thinking mindset to ideate *radically*. What this means for UX writers is to facilitate brainstorming that leads to fluency (volume) and flexibility (variety) in their innovation options. Radicality also means thinking beyond technical/functional and material constraints. It is not designing for designing sake or merely seeking to create a shinier or faster version of an existing product. Radical ideation aims to challenge conventional assumptions about people, society, and things. It also aims to advocate for justice and equity for those who have been traditionally oppressed by the current design. Radical ideation can lead writers to uncover unexpected design solutions. But because radical solutions are often unconventional and thus could cause discomfort, you should mentally prepare yourself to expect this discomfort.

Last, ideation benefits from diverse perspectives and opinions. So, in the following sections, we provide tips for bringing in people you don't usually work with or speak to so you can listen to what they have to say. Collaborating with everyday users during the ideation process can lead to more relevant solutions. Let's begin with card sorting.

What Is Card Sorting?

Ideation is an active learning and generative process. It is not just a group of writers sitting at a table listing potential ideas. Card sorting is a popular user-centered design method that allows writers to "see" how their users think or make sense of content information. Essentially, you invite a few participants who are representative of your audience population to take part in an information-sorting exercise. Participants sort concepts, terms, or features into meaningful categories based on relationships that they create.

Card sorting is an inexpensive yet powerful way to understand users' mental models. By observing how participants group things and listening to their conversations as they do so, writers can identify terms that might be confusing, easily misunderstood, or too complicated for the target audience. As well, card sorting reveals how people navigate and structure information, which can inform the design of interface organization, menus, and taxonomies. There are two types of card sorting: open and closed sorting.

Open card sorting asks the participants to organize the cards presented to them, sorting them into relevant categories, and creating appropriate labels for each category (Figure 6.2). UX writers use this method to ideate information architecture, especially when the content interface is new or unfamiliar to everyday users. Rather than a top-down design approach, open card sorting is a bottom-up approach that may reveal interesting user cognitive behaviors.

Closed card sorting begins with the same protocol as open sorting, where participants are presented with cards to sort into categories. The difference here is that participants do not get to create their own labels for the categories. The writers give participants a predetermined set of labels and ask them to sort cards into these labeled categories. The purpose of this approach is to evaluate existing structures. These existing labels typically come from the results of content auditing (see Chapter 5). Close card sorting can reveal user impressions toward the given taxonomies and show writers how effective or ineffective the previous design is (Figure 6.3).

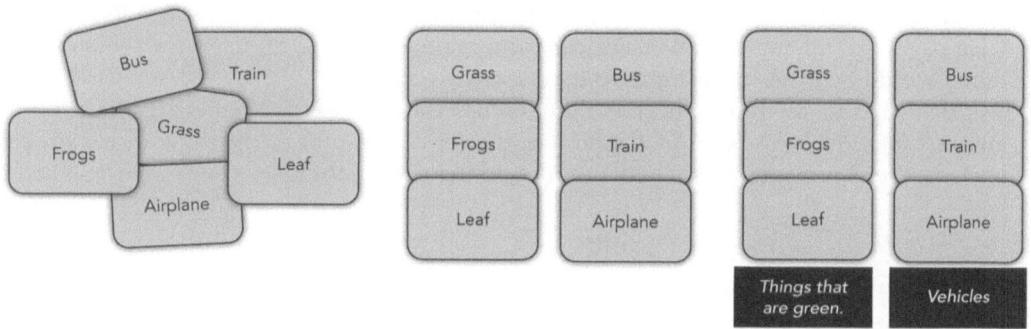

1. Participants get a stack of cards.   2. Participants sort cards into groups.   3. Participants label the groups.

Figure 6.2 Open card sorting.

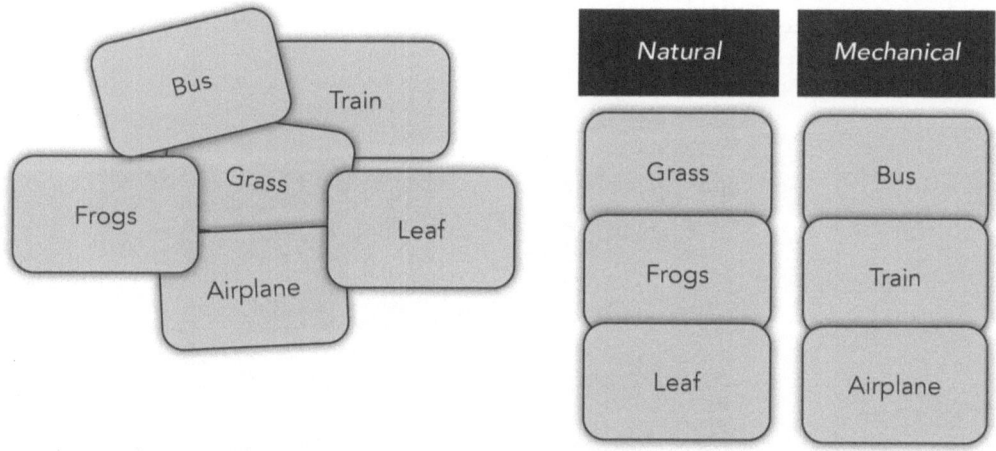

1. Participants get a stack of cards.

2. Participants sort cards into groups the researchers have created.

Figure 6.3 Closed card sorting.

A hybrid approach is possible if the writers deem it necessary. In hybrid card sorting, participants get to sort cards into the categories you provide as well as create their own labels for cards that don't fit into any existing labels.

If you are using card sorting in ideating content solutions, consider the following tips:

- Recruit 10–15 participants to make enough groups for open, closed, or hybrid card sorting. Say if you want five groups of three, you need to recruit 15 participants.
- Consider the cognitive load on the participants. Limit topics to individual webpages, menus, labels, or product features in somewhere around four to five categories.
- Pretest the topics with your team or other UX writers to make sure the cards are indeed sortable.

- Avoid synonyms on cards. They confuse participants and mess up your results.
- Give participants enough time to discuss and make decisions. Somewhere around 1–1.5 hours would be ideal for a standard card sorting session.
- Ask participants to think aloud as they discuss their decisions. Similar to the protocol analysis method described in Chapter 4 and later in Chapter 7, the think-aloud process lets you "see" how the participants think and make decisions.

What Is Affinity Diagramming?

When UX writers perform an activity similar to card sorting, it is known as affinity diagramming. It is an ideation exercise that uses findings from the research phase to create meaningful categories or relationships—hence, "affinity" —that tell writers what is currently happening and what could be done in the future. Pick up any UX book, and you will likely see this signature activity represented by an image that shows a wall of sticky notes (like Figures 1.1, 4.1, and now 6.4). It's almost as though affinity diagrams are the face of UX design. At any rate, affinity diagramming is a rigorous way to generate new ideas based on actual data.

Figure 6.4 A wall full of affinity categories on sticky notes.

Source: Hugo Rocha on Unsplash.

How do you perform this activity, exactly? Follow these steps:

1 Locate a room and a large surface (like a wall, whiteboard, table, etc.) where you could host multiple people meeting and creating affinity diagrams.
2 Gather all data and findings from your research, such as contextual inquiry, interviews, focus groups, and surveys.
3 Use colored sticky notes to create visual categories.
4 Write one significant result, observation, issue, or idea per sticky note.
5 Ask the writers to sort all the sticky notes into relevant categories. When sorting, record the conversations the writers have and the questions they raise.
6 When the categories are ready, write labels that represent each category's characteristics with short descriptions.
7 Review the affinity categories as a whole. Identify overlaps, distinctions, outliers, or extreme cases. Actively discuss any agreement or disagreement on the categories.
8 Create new categories or write new descriptions based on the discussion.
9 Report on the insights gained from the exercise.
10 Optional: Digitize the affinity diagrams so they can be preserved for future reference.

After the diagramming activity, writers should brainstorm ideas for issues in the product that could be addressed by design. At this stage, writers usually branch out into different discussion points to consider viable directions. Remember, do not evaluate ideas just yet. Think wide and account for all suggestions at this time.

What Is Participatory Design?

Since the late 1960s, North American designers have been fascinated by a then-new development in design practice popularized by Scandinavian practitioners who showed how everyday users can illuminate the invention process. In particular, American designers were intrigued by how Scandinavian designers invented creative but, more importantly, practical design solutions simply by involving users in the design process. Originally called the co-design method, this practice was soon adopted by entrepreneurs and innovators who sought to meet user-centered design standards. In the U.S., prominent designers called this approach participatory design—a way of innovation that engages actual users throughout the process of generating ideas, mocking up prototypes, and testing beta products.

The participatory design method makes for a significant part of the ideation phase as it affords writers the opportunity to get inspiration from novice users. It helps writers observe mundane user behaviors that are usually unavailable to them due to the way content is used at work, at home, or in other personal contexts. It respects the creative insights of participants to help guide the design process and respond to design outcomes. It also creates an open channel for non-designers to voice their hopes, needs, and wants for a product. For the purpose of ideation, participatory design typically involves activities that ask participants to role play, tackle certain scenarios, and co-discover solutions.

UX writers can facilitate participatory design workshops by recruiting representative participants to spend a few hours (or a few days) interacting with one another to complete the exercises given by the writers. In a role play session, participants are asked to act out an assigned role to perform a variety of tasks. The tasks and scenarios are intentionally designed to let participants experience a product or its prototype. In doing the exercise, participants typically talk among themselves to solve the given problem. This interaction,

Figure 6.5 Participatory design activity featuring collaborative sketching.
Source: Amélie Mourichon on Unsplash.

known as a co-discovery protocol, allows the workshop facilitators to identify participants' problem-solving processes, mental references, and social actions, all of which are invaluable for inspiring new ideas.

Figure 6.5 shows an example of a participatory design activity where participants sketch mobile interfaces together. In some workshops, participants are also given modular design tools (like Legos, papers, and sticky notes) to invent a version of the design they envision. This can spark interesting and important conversations between participants and the writers.

What Are the 6:1 and Four-Category Methods?

It is quite common for writers to retreat into more rational design ideas during ideation sessions. The 6:1 method is a common trick to help writers come up with radical rather than rational ideas. Tell yourself you may ideate one rational solution *only after* you have come up with at least six radical ones. The radical ideas should be bizarre, impractical, and idealistic. Studies have shown that applying time constraints to the brainstorming process can help squeeze the creative juice in your head. So, perhaps you may practice timing yourself (two to five minutes) to come up with as many ideas as possible. Once you have six radical ideas, you may ideate a rational one to serve as a benchmark for the solution. Then hybridize the ideas. Combine all radical and rational solutions into one semi-practical solution. You may find this hybrid solution to be a good starting point for prototyping later.

The next step is to use the four-category method to visually order your ideas. This method is meant to give writers a chance to discuss their favorite ideas and how amenable they are. The four categories are:

- The long shot: Ideas that are most idealistic and impractical.
- The darling: Ideas that are the most favored by the team.
- The most delightful: Ideas that would most likely make the user happy.
- The most rational: Ideas that are grounded in a pragmatic or practical sense.

Just like card sorting and affinity diagramming, organize the ideas that were generated during ideation sessions into these four respective categories. There's no limit on how many ideas can go into each category. Once all the ideas have been sorted, spend time talking about them with your team. This is usually the time to select an idea to move forward. Teams can discuss and debate which of these categories and the ideas they contain cover the best ground for addressing the problem at hand.

Selecting a Solution

The four-category method is useful for selecting solutions that will be built and tested. The mindset to have at this transition stage is convergence. Instead of wanting to accomplish every great thing that came up during ideation, writers need to focus on detailing the concepts from a general to a granular level so the team can select the best concept(s) to move into prototyping.

Pugh's (1990) model for concept selection, adopted by Buxton (2007), demonstrates a desirable "funneling down" manner that keeps writers focused on choosing one thing to build at a time (see Figure 6.6). Depending on your project, you may need to consider the scope of this "one thing" —is it a form? A webpage? A website? A help forum? A full user guide? An employee training module? These materials vary in length and complexity and should be considered deliberately by your team.

Prototyping: Materializing Ideas

After you have selected the best concept or solution to move forward, the next step is to materialize the idea into a physical (or digital) reality. This phase, known as prototyping, is about building tangible content to allow actual users to interact with it. The goal of prototyping is to bring an idea out of your head and into material form in the real world. The two defining characteristics of a good prototype are that it should be *tangible* and *testable*. Tangibility can be achieved when the idea you generated is given life through a 2D or 3D treatment. It is when an idea becomes a thing. Although a prototype doesn't have to be a physical manifestation, and in fact, a lot of UX writing can be prototyped digitally, it needs to be concrete enough to be testable. A testable prototype is one that a user can interact with and know what the interaction represents. For instance, hand sketches can be tangible (a user can hold the sketch board in hand), but they may not be testable. To become testable, the designers may need to create paper constructions to represent the elements in the sketches so that the user can interact with those elements.

CONCEPT GENERATION

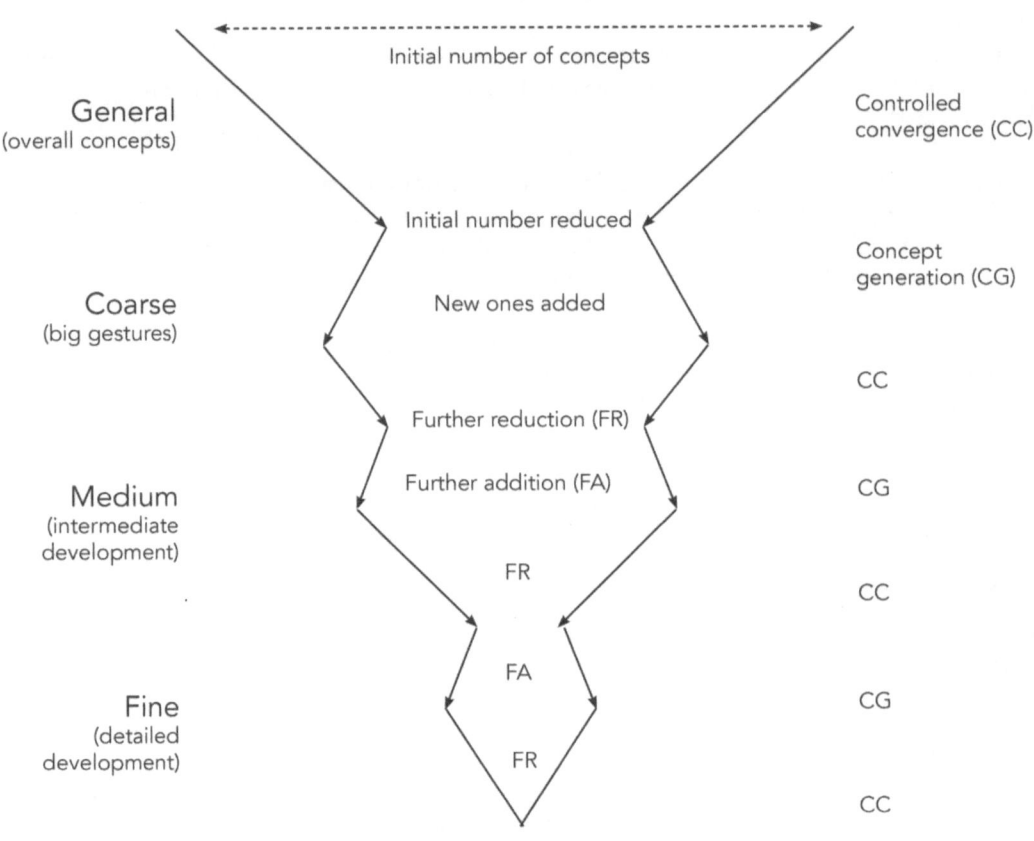

CONCEPT SELECTION

Figure 6.6 The convergence process before prototyping.

Source: Adapted from Buxton (2007).

Prototypes are meant to bring your ideas to the real world and show your intended users and stakeholders how they can interact with your designed solution. Prototypes naturally contain rough edges and placeholder content. You should not worry about perfecting your model but instead focus your energy on giving your ideas a material form. Just start building. Prototyping helps you think about your ideas in a concrete manner and gain insights from testers about ways you could improve them.

When building your prototype, keep your users in mind at all times. Go back to the personas and associated user journeys you have previously generated to guide your prototyping process. If your participatory design workshop participants are willing to maintain contact with you, leverage their availability to gather ongoing feedback throughout the prototyping phase.

While the testing phase is yet to come in the next stage per design thinking workflow, you may share your early prototypes with colleagues and friends to get early feedback. This kind of informal testing may help you avoid glaring issues in your design during formal testing later. All prototypes should have a central testing issue. It is important to not lose sight of that issue, but at the same time, you should not get so bound to it as to lose sight of other lessons you could learn from prototype users. When you receive feedback from coworkers and friends about your ongoing prototype, use the feedback to iterate your design. However, don't spend too much time trying to get everything right by addressing every single concern in the user feedback. Prototyping is about *speed*; the longer writers spend building a prototype, the more emotionally attached they could get to that idea, thus hampering their ability to objectively judge its merits. Based on our experience, our advice is to just build, share, and reflect, and then move on to formal testing in the next phase.

Indeed, prototyping might sound daunting to some writers who are not comfortable with design applications. Worry not, you may have already created pieces of your selected concept here and there when performing ideation activities such as card sorting, the 6:1 method, and participatory design workshops. Look for sketches made on paper and reflective notes written on sticky notes. These are great starting points for your prototype, as they are useful for creating lo-fi prototypes that can then be refined into hi-fi prototypes for testing later.

What Is Low-Fidelity Prototyping?

Lo-fi prototyping involves the use of basic models or examples of the product being tested. For example, the model might be incomplete and utilize just a few of the features that will be available in the final design, or it might be constructed using tools and materials not intended for the finished product, such as papers and sticky notes, cardboards, whiteboards, digital canvases, and placeholder content (see Figure 6.7).

Lo-fi prototypes are quick and inexpensive to build. They allow for instant modifications and the testing of new iterations of ideas. As shown in Figure 6.7, paper prototyping is the most common lo-fi prototyping method. When building a lo-fi prototype, details are not important. Writers should focus on creating a skeleton structure for their concept, outline key content, and mark special features or design elements that help the team and client see an "operative image" (Still & Crane, 2017) of the concept you're building.

When creating lo-fi prototypes, define your goals and use scenarios so you can apply them in testing sessions with users, like in the following example.

Goals:

- Understand the user's navigational behavior with "pages" on wearable interfaces.
- Find out the user's mental model for browsing a website on a smartwatch.
- Explore the user's comfort level with smaller-screen interfaces.

Use Scenario:

You're an active college student who likes to find information quickly on mobile devices. Just the other day, you were wondering how to air-fry chicken wings, and you found your answers with just a quick search on your phone. Last week, you were given an Apple Watch for your birthday. You learned that the watch is connected to the internet and now wonder how you can search for online content using your watch. We have prototyped a few "mini webpages" that can perform basic browsing. Show us how you would use these pages to find what you need. If you could design your own browsing feature, what would it look like?

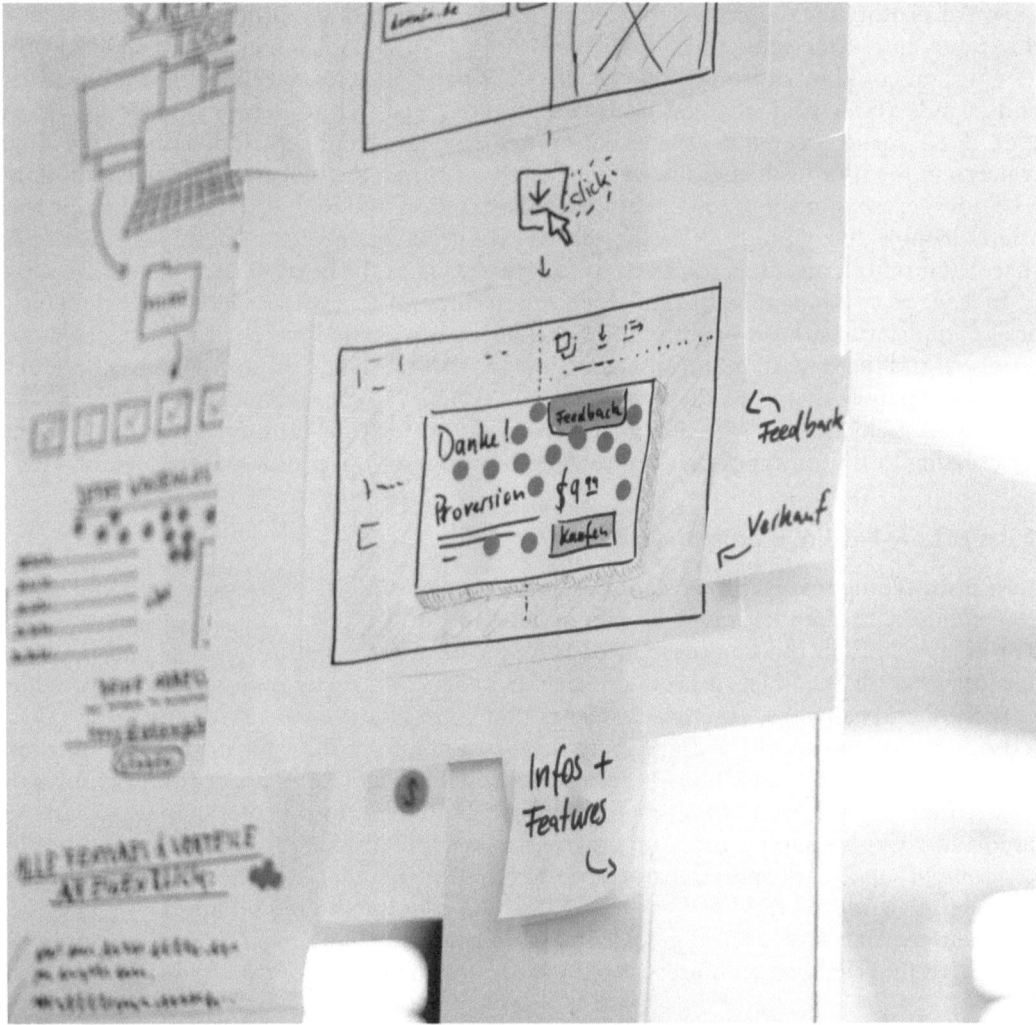

Figure 6.7 Low-fidelity prototyping uses modular pieces to create temporary models that can be rearranged if desired.

Source: New Data Services on Unsplash.

What Is High-Fidelity Prototyping?

Hi-fi prototypes are prototypes that look and operate closer to the finished product. An early version of a software system developed using a design program such as Adobe Illustrator has hi-fi in comparison to a paper prototype. Hi-fi prototypes are more engaging; they allow the users and stakeholders to see their vision realized and be able to judge how well it meets their expectations, wants, and needs. User experience testing involving hi-fi prototypes allows the evaluators to gather information with a higher level of validity and application compared to lo-fi prototypes.

To build a hi-fi prototype, start by collating the feedback you received from people who have interacted with your lo-fi prototype. What did they like about the design? What did they recommend to change? What modifications are you and your team going to make to the selected concept? What will you retain in the design? Using these findings, generate a list of design iterations and make decisions for your hi-fi prototype.

Then, you need to find an appropriate tool or platform to create your interactive prototype. In Table 6.1, you will see a selection of applications that can be used for print or graphic design, digital mock-ups, and website development. Print design software lets you present the visual interface of your content with artistic details such as colors, shapes and objects, typography, and other page effects. Mock-ups are realistic renderings of your design that can contain some interactivity (such as buttons, links, and animations). Website builders allow you to house the actual content you wish to display without the need to worry about coding and programming at this stage.

Needless to say, hi-fi prototypes are more costly and take longer to produce. The trade-off is the experience your test participants may get from interacting with a more polished prototype, which can affect the quality of their feedback on the design. The more resembling the prototype is to the actual product in appearance, look and feel, functionality, and interactivity, the more representative the user experience will be of the actual experience.

However, for teams that do not have the team or immediate resources to create a dynamic hi-fi prototype, they can consider what Jeff Gothelf (2013) called a minimum viable product (MVP; also see Chapter 4 for details). For digital interfaces, an MVP could be a wireframe (like Figure 6.8). Sometimes called mid-fidelity prototyping, wireframing involves using lines and basic shapes to represent the desired content on a page without providing detailed information about it. Designers use them to communicate the arrangement and overall layout of the content, skipping stylistic details like colors, typography, and images. They are typically static (non-reactive to clicking or scrolling) and without animation. While, of course, hi-fi prototypes can be built with design software to give reviewers a better understanding of your vision for content development, wireframes can be built faster and easier, thus making them more desirable for projects that need a quick turnaround in content design and testing.

Table 6.1 Tools to create hi-fi prototypes

Graphic/print design	Adobe Creative Suite (Illustrator, InDesign, and Photoshop)	https://www.adobe.com/creativecloud.html
	Canva	https://www.canva.com/
	QuarkXPress	https://www.quark.com/
Interactive mock-ups	Balsamiq	https://balsamiq.com/
	InVision	https://www.invisionapp.com/
	Proto.io	https://proto.io/
	Axure	https://www.axure.com/
	Adobe XD	https://www.adobe.com/products/xd.html
Web building	Wix	https://www.wix.com/
	Squarespace	https://www.squarespace.com/
	WordPress	https://wordpress.com/
	HTML5 Up	https://html5up.net/

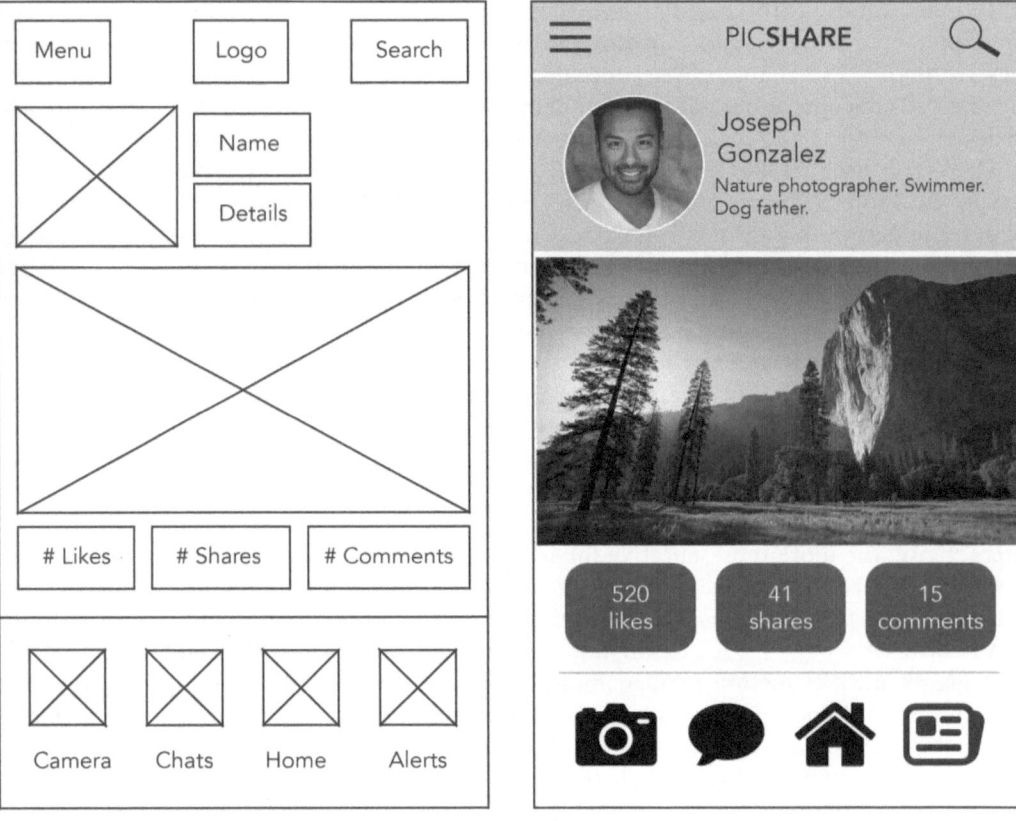

Wireframe   **Mock-up**

Figure 6.8 A wireframe (left) vs. mock-up (right).

Preparing to Test

When do you know if a prototype is ready? Just remember that a prototype is only a proof of concept, and it's going to have rough edges. A ready-to-use prototype should manifest your idea in material form. The best rule of thumb to go by when judging if your prototype is ready is to ask the question: Is it testable? Specifically, does it have the following qualities (adapted from RevPart, n.d.):

- Illustrates the real-world functionality of the content or product you aim to deliver.
- Shows how the content fits and interacts with other existing components of a design.
- Provides a close facsimile of the intended aesthetics (or shows an indication of what the end product should look like).
- Demonstrates that the content can, in fact, be produced within available resources and budgets.

Once you've determined that your prototype meets these criteria, you should move forward with planning different kinds of testing with real users.

Conclusion

Ideation in UX writing is biased toward divergent thinking. Based on the insights gathered from your research on users, contexts, and issues, you can develop rational as well as radical solutions to address specific design problems. Popular methods to help teams brainstorm ideas involve users in the process, such as card sorting and participatory design workshops. Writers can use affinity diagramming and the 6:1 and four-category methods to select the most appropriate concept. Using cheap or readily available materials, you can create lo-fi prototypes, such as paper prototypes, to use in informal testing. When ready, these lo-fi prototypes can be built into mid- or hi-fi prototypes using design applications to enhance interactivity and resemblance to the actual content. These prototypes will be used in the formal testing sessions we discuss in the next chapter.

Chapter Checklist

This chapter has covered two very important phases of UX writing, where you were introduced to major methods for generating and then building UX content.

- *Practice divergent thinking when ideating.*
  - Use the defined problem to guide your ideation process.
  - Avoid groupthink and do not settle on early or easy solutions.
- *Use card sorting to understand users' mental models.*
  - Use open card sorting to explore user cognitive behaviors.
  - Use closed card sorting to examine users' impressions toward given taxonomies.
- *Perform affinity diagramming to identify patterns.*
  - Create meaningful categories or relationships from your gathered findings.
  - Discuss viable directions for solutions based on emergent themes.
- *Facilitate participatory design workshops to get user input.*
  - Involve actual users in the ideation process with participatory design workshops.
  - Observe and record how participants interact with the assigned tasks and with each other.
- *Use the 6:1 method to encourage radical innovation.*
  - Allow yourself one rational design idea only after coming up with six bizarre solutions.
  - Hybridize radical and rational ideas to create innovative yet practical solutions.
- *Use the four-category method to discuss and select ideas with your team.*
  - Visually represent and order your available solutions.
  - Reach consensus on the selected concept based on discussion.
- *Materialize your chosen concept with prototyping exercises.*
  - Put ideas into the world via physical or digital prototyping.
  - Build, share, and reflect with prototypes.
- *Create lo-fi prototypes for initial testing.*
  - Prototype quickly with cheap and available materials like papers and sticky notes.
  - Test early and often with lo-fi prototypes.

- *Build hi-fi prototypes to enhance evaluators' experiences.*
  - Use design software/applications to create hi-fi prototypes.
  - Create mid-fidelity, or MVP, for a faster turnaround in feedback and testing.

Discussion Questions

1 What is the difference between divergence and convergence in the design process?
2 How does card sorting afford an understanding of user mental models?
3 Describe the things you should pay attention to when facilitating a participatory design workshop.
4 What's the difference between lo-fi and hi-fi prototypes?

Learning Activity

Let's try building a lo-fi and then a hi-fi prototype for this scenario: Customers at Amarillo Happy Bank (AHB) have experienced confusion in understanding their credit scores that were made available to them via the bank's mobile app interface. The app's current page only shows a number of AHB customers' scores against the total possible score. AHB customers do not know how to interpret their scores or how to improve them. You are asked to create an interface with user-centered content that helps AHB customers understand their credit scores.

Step 1: Sketch out, on a piece of paper, a few potential interfaces and copy that could better inform AHB customers about their current scores. Consider how people read credit scores, how they understand credit standing, and how credit might be important to them.

Step 2: Select one of the available digital tools (see Table 6.1) to build a potential interface after speaking to a few individuals who use digital apps to view their credit scores. Develop interactive (or clickable) wireframes for your design that allow for informal testing to identify what seems intuitive and what appears to be confusing.

Step 3: Document your process and feedback. Write a reflective journal entry about your design decisions and process, and be candid about your choices. Then, upon receiving feedback from potential users, write down your reactions to the feedback and how or why you might be using that feedback to improve your design.

References

Buxton, W. (2007). *Sketching user experiences: Getting the design right and the right design.* Morgan Kaufmann.

Gothelf, J. (2013). *Lean UX: Applying lean principles to improve user experience.* O'Reilly Media.

Janis, I. (1991). Groupthink. In E. Griffin (Ed.), *A first look at communication theory* (pp. 235–246). McGrawHill.

Pugh, S. (1990). *Total design: Integrated methods for successful product engineering.* Addison-Wesley.

RevPart. (n.d.). 8 keys to a successful prototype. https://revpart.com/keys-to-successful-prototype/

Still, B. & Crane, K. (2017). *Fundamentals of user-centered design: A practical approach.* CRC Press.

von Oech, R. (2021). *The creative contrarian: 20 "wise fool" strategies to boost creativity and curb groupthink.* Wiley.

7 Testing, Managing, and Deploying Content

Chapter Overview

This chapter covers the steps in testing and validating content using the materials created during the ideation and prototyping phases. This important stage of the UX writing process seeks to gather critical feedback from representative users to create polished content that will be deployed into the world. The chapter will also discuss the benefits of structured authoring methods and the use of content management systems (CMSs) to design omnichannel content capable of reaching the user at their point of need in their preferred medium.

Learning Objectives

- Describe the stages of testing, managing, and deploying content.
- Conduct usability testing, A/B testing, and heuristic evaluation appropriately.
- Practice structured authoring methods to optimize UX writing.
- Use CMSs to organize and manage developed content.
- Deploy omnichannel strategies to deploy content effectively across multiple channels.

Testing and Validating Content

Once you and your team have settled on a prototype that has received favorable responses from informal trials, the next step is key in the UX writing process. Using the hi-fi prototypes you have created, you will perform formal usability testing to capture realistic user performance and validate the capability of the content to achieve desired goals. Testing is the only way you can find out whether your design has accomplished the following quality attributes:

- **Accuracy of information:** Your content is factually accurate, consists of the information users seek, and is up to date.
- **Ease of learning:** Content is designed for quick comprehension, avoids technical jargon and high-level concepts (unless it's meant for an expert audience), and is easy to remember.

DOI: 10.4324/9781003274414-9

- **Efficiency of use:** Content is accessible, doesn't require a lot of time or steps to understand or apply, and is relatively effortless to use.
- **Effectiveness of performance:** Content supports various user activities, provides support where it's necessary, and helps users avoid committing errors.
- **Satisfaction of users:** Content is desirable and makes users feel positive about using it.

Although you've been testing the design of your content iteratively throughout the UX writing process—from empathy research to defining user and content requirements to ideating and prototyping viable solutions—you have likely done so internally with your colleagues or people immediately accessible to you (like a family member or a friend). Formative testing requires you to get feedback externally, specifically from users representative of the target audience you're designing for. In the next sections, we will go over the different methods for testing with real-world participants.

What Is Usability Testing of Prototypes?

Usability testing (or what we called "protocol analysis" in Chapter 4) is the cornerstone of user-centered design. In Chapter 4, we discussed how think-aloud protocols can be used to help you develop empathy for your users, but in this chapter, we want to show how they can be used for testing and evaluation. It is the deliberate evaluation of a product or service to determine its applicability and usefulness by inviting real-world users to examine it. Previously, we discussed ways usability testing is conducted in controlled environments, such as a testing lab, an office, or a conference room. The purpose of doing the tests in an isolated space is to allow the participants to focus and perform the tasks instructed by the test facilitator using the prototyped product or content.

However, usability testing can be performed in person or remotely, moderated or unmoderated. In-person testing is one where the researcher or facilitator and the participant are present in the same room at the same time and interact with each other. Remotely moderated testing is done when the facilitator and the participant are in different places but are iterating synchronously using conferencing technology. Remotely unmoderated testing is when the researcher isn't involved while the participant does the given tasks at a time and place of their own choosing. Moderated usability testing allows for control over the test session, adaptation of tasks based on the participant's performance or reactions, and is relatively cheaper than unmoderated testing. Comparatively, unmoderated testing offers flexibility by letting participants perform the test in their own comfortable environment and is arguably more realistic because the participants are being watched by a facilitator or observer. Unmoderated testing also gives you insights into how participants solve problems independently without facilitator intervention.

The first step to conducting a usability test is to create a script (also known as a protocol) with the scope of the testing session, the particular scenarios and associated tasks for testing a product, and the rundown of the test (what happens first, next, and last). Before you start recruiting people to participate in your test, prepare the necessary informed consent materials that specify the rights of the participants. In the consent form, you need to let individuals know:

- What they will do?
- What personal data you will collect?
- How you will record their performance (written notes, screen capture, audio, or video)?

- What you will do with the data?
- What you will do with the data after your project is complete?
- How will you ensure that all personal or identifiable information is protected?
- What risks and incentives will the participants receive?
- How they can withdraw from the test session at any time?

You may recruit participants using a recruiting agency or web service, or may use an open enrollment method to attract participants. In your recruitment process, you need to specify basic information about your target audience, including their demographics, experience, expertise/skill sets, education or work experience, etc. This can help you locate and gather people who best suit the needs of your testing.

Real World Snapshot 7.1: How Many Participants Is Enough for Usability Testing?

Just like any other proper research method, usability testing aims to achieve rigor through robust design and a process that yields valid and reliable results. But every human-subject researcher has to deal with the century-old question of sample size, as with usability studies: How many participants should I recruit for my usability test?

The federal resource, Usability.gov, recommends just five participants for a standard usability test. This number is not arbitrary. The magic number is derived from one of Jakob Nielsen's influential studies about the significance of quantity versus quality. In "Discount Usability Engineering," Nielsen (1989) demonstrates how simplified thinking-aloud and heuristic evaluation are sufficient for locating major problems in product testing. Robert Virzi (1992) created a model based on other usability projects, finding the p-value between 0.32 and 0.42. Therefore, 80% of the usability problems in a test could be detected with four or five participants. Later, Nielsen and Thomas Landauer (1993) showed that the number of usability problems found in a usability test with n users is:

$$N(1 - (1 - L)^n)$$

where N is the total number of usability problems in the design and L is the proportion of usability problems discovered while testing a single participant. The typical value of L is 31%, averaged across a large number of projects studied. Plotting the curve for $L = 31\%$ gives the following result (Figure 7.1).

By this formula, the third test participant will do many things that you already observed with the first or second participant, and even some things that you have already seen twice. According to Nielsen and Landauer, after the fifth participant, there wouldn't be many new things to observe.

Of course, we need to consider the context and goals of every study. If your objective is to locate statistical significance in your results, the Nielsen Norman Group recommends at least 20 participants for quantitative approaches.

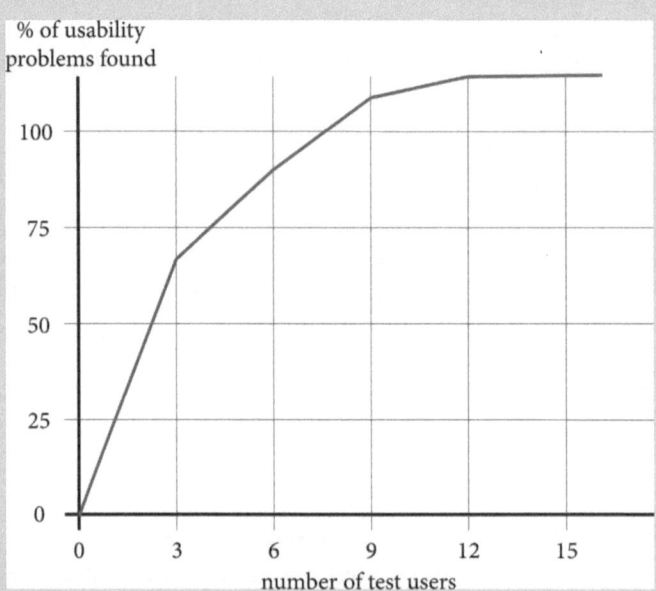

Figure 7.1 Number of test users to find the number of usability problems according to Nielsen and Landauer's (1993) model, adapted by authors.

After you have gathered the desired number of participants, you can then schedule and invite the registrants to come to your testing location at a given time and date. To ensure a low dropout rate, be sure to remind your participants a day or two before the session and give them a chance to ask questions about logistics.

When your participant arrives at the designated location, begin the test session with a pre-task questionnaire or conversation. The purpose of this exercise is to gather any pre-exposure feelings, attitudes, or conceptions the participant may have about the product with which they will interact. This information may reveal telling insights when compared with post-task reactions after the test. When ready, get the participant situated at the test station, making sure that they are comfortable and that all accommodation needs have been met (e.g., visual, auditory, or motor/physical assistive setups).

Give participants simple and straightforward directions on what to do with the product presented to them. Remind them that the objective of the test is to evaluate the product, not the participant. Then, teach them how to "think-aloud" as they interact with the product (see Chapter 4). Thinking aloud means talking out loud the thoughts, feelings, or questions that emerge during the interaction. This method can be deemed awkward by many people, so the facilitator needs to always gently remind the participants to verbalize their process if they become silent.

When conducting the test, be sure to tell the participant the scenario for the tasks they are performing with the tested product. A scenario is the backdrop that sets the scene for the activity. Tasks are the step-by-step procedures for accomplishing a goal (like buying something in an online store using a mobile app). Below is an example that may be familiar to many students.

If you are working solo, you will play the role of facilitator as well as observer of the participant's process in completing the assigned tasks. If you have the luxury of working in a team, you may leverage the availability of other team members by asking them to observe from an observation room and document what they see and hear. In some testing, eye-tracking and mouse-tracking methods are used to record how the participant navigates spaces/pages on a screen and what they fixate on. Look back to Chapter 4 for more discussions on eye-tracking.

Scenario: You are a college student who wants to cite three academic sources for your argumentative essay assignment. You have decided to use the University Library website to locate at least three journal articles to support your claims.

Tasks:
1. Starting at the library homepage, locate the OneSearch bar.
2. Find appropriate citation information in the returned results.
3. Download the full text of each selected article.

When the participant is done with the tasks, the facilitator then closes the session with a short interview with the participant to ask some post-task questions. The purpose of this practice is to gather reflective responses on the test experience—anything surprising to the participant, overall ease of difficulty of use, top issues, emotive response, etc. —that would allow you to complement the quantitative results from the test. It is recommended that you meet briefly with your team immediately following a test session to collate your observations so you don't forget them before your formal analysis meeting. Discuss what top positives and negatives have been seen in the participant's interaction with the product, and write down specific verbalizations of the participant that you'll go back to extract from your recordings later.

After you have completed the desired number of test sessions with your team, you can analyze the observations and results from these sessions. There are two categories of findings (Table 7.1).

Quantitative results are those with numerical values that you can measure and compare among participants. Task completion rates are the frequency and percentages of success versus failure in each participant's interaction with the product. Error rates are the number of missteps committed by the participants, with an assigned rating of how severe each error is. Low severity means problems that could be easily avoided or errors from which the participants recovered quickly. High severity is assigned to errors that were showstopping—took the participants off course, and could not easily be recovered from. Other quantifiable measures include the amount of time it takes for the participant to complete a task, how long they spend on individual pages, and the number of clicks done per task. These numbers can be used to generate revision ideas to optimize the efficiency of the content design.

Qualitative findings come from personal reflections and ratings by the participants during and after they did the tasks. During the test, you can identify verbal (spoken) as well as nonverbal expressions by the participants that indicate satisfaction or frustration with the process. During the post-task interview, you can administer standardized questionnaires like the

Table 7.1 Types of results from usability tests

Quantitative	Qualitative
• Task completion rates	• System usability scale
• Error rates & severity	• Satisfaction scale
• Time on task	• Interview responses
• Time on pages	• Nonverbal expressions
• Clicks per task	

System Usability Scale (SUS; see template here: https://www.usability.gov/how-to-and-tools/methods/system-usability-scale.html) and NASA's Task Load Index (TLX; https://humansystems.arc.nasa.gov/groups/tlx/tlxpaperpencil.php) to gather participants' ratings for ease and difficulty in the product use. As indicated earlier, direct or verbatim quotes by the participants are also valuable, qualitative results from usability testing (Figures 7.1 and 7.3).

1. I think that I would like to use this system frequently.

1. Strongly Disagree	2.	3.	4.	5. Strongly Agree
○	○	○	○	○

2. I found the system unnecessarily complex.

1. Strongly Disagree	2.	3.	4.	5. Strongly Agree
○	○	○	○	○

3. I thought the system was easy to use.

1. Strongly Disagree	2.	3.	4.	5. Strongly Agree
○	○	○	○	○

4. I think that I would need the support of a technical person to be able to use this system.

1. Strongly Disagree	2.	3.	4.	5. Strongly Agree
○	○	○	○	○

5. I found the various functions in this system were well integrated.

1. Strongly Disagree	2.	3.	4.	5. Strongly Agree
○	○	○	○	○

6. I thought there was too much inconsistency in this system.

1. Strongly Disagree	2.	3.	4.	5. Strongly Agree
○	○	○	○	○

7. I would imagine that most people would learn to use this system very quickly.

1. Strongly Disagree	2.	3.	4.	5. Strongly Agree
○	○	○	○	○

8. I found the system very cumbersome to use.

1. Strongly Disagree	2.	3.	4.	5. Strongly Agree
○	○	○	○	○

9. I felt very confident using the system.

1. Strongly Disagree	2.	3.	4.	5. Strongly Agree
○	○	○	○	○

10. I needed to learn a lot of things before I could get going with this system.

1. Strongly Disagree	2.	3.	4.	5. Strongly Agree
○	○	○	○	○

Figure 7.2 The SUS contains ten different questions that address the usability and learnability of a system. Do not alter the order or the wording of the SUS questions if you want to compare your score with the scores collected from other designs.

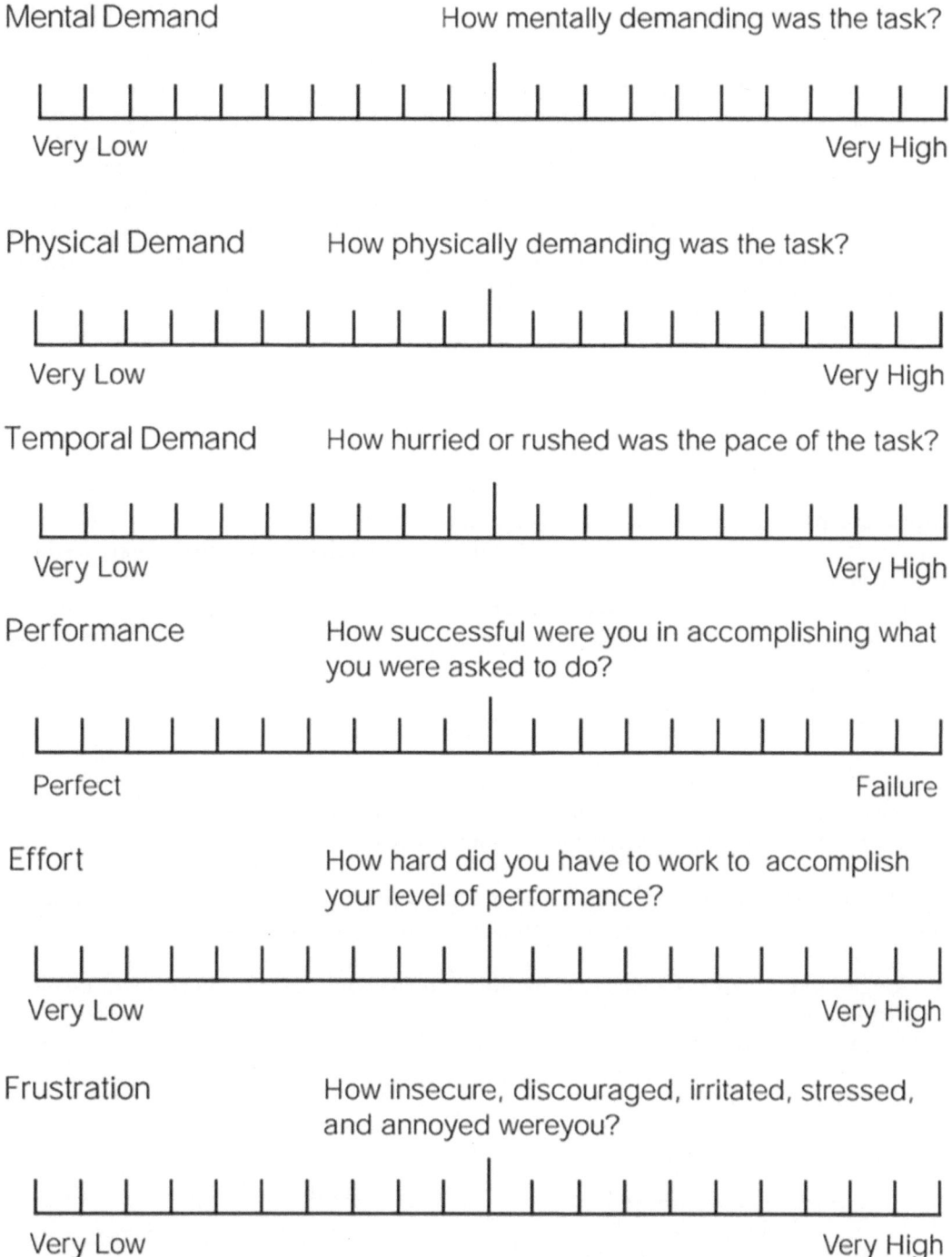

Figure 7.3 The NASA-TLX instrument asks participants to rate each task they have performed on these six scales, each in 21-point increments. It provides rich data about what sorts of demands the task has on the user in multiple different areas, but it requires time and expertise to collect during a study.

Limitations of Usability Testing

It is important to keep in mind that usability testing is limited by several factors. Due to the sample size, the results from usability tests are not statistically significant, although they can be representative of the general user experience. During test sessions, moderated or not, user performance can be unnatural, thus leading to inauthentic responses and results. As mentioned earlier as well, it is also difficult to keep participants "thinking aloud" because it is not common to people. And finally, the dropout or no-show rate is typically high unless the incentives are lucrative. For these limitations, you should plan to triangulate your testing methods by relying not just on usability testing but also on other sources of participant reactions to your product design, such as A/B testing and expert review (also called heuristic evaluation).

What Is A/B Testing?

Commonly applied to web-based and mobile application interface design, A/B testing is, simply put, a systematic way to compare two or more versions of content (A, B, or C, etc.) to determine which version receives better feedback. It also allows you to see if the difference in comparison is statistically significant (more on this in a bit). UX writers conduct A/B testing after content has been prototyped and do so in conjunction with usability testing to reach a greater number of respondents.

To run an A/B test, you first need to determine the purpose of the test. What is it that you want to learn? Did you want to know if the title or heading of the content affects reading speed? Or if different call-to-action buttons lead to differences in conversion rate? What about the color scheme and its effects on bounce rate (i.e., the number of visitors leaving a page without taking an action, like clicking on a link or filling out a form)? Once you have identified your testing goals, the next step is to create an alternative version of the prototyped content. It is best to keep differences parallel—i.e., apple to apple: compare header to header and button to button. If you mix elements of comparison, you may not be able to tell if a difference in action was indeed caused by one or more of the differences (Figure 7.4).

When your variate versions are ready to go, use a test management tool to administer the test to selected users. A few easy-to-learn tools we recommend are:

- Google Optimize (https://marketingplatform.google.com/about/optimize/features/)—free
- VWO (https://vwo.com/)—paid
- Optimizely (https://www.optimizely.com/)—paid

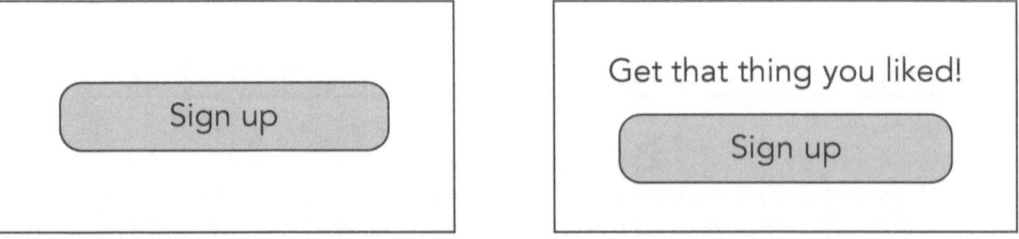

Figure 7.4 Variations in the design of similar content.

These tools allow you to perform split URL testing, get real-time responses from robust reporting dashboards, and run multiple-variable (multivariate) experiments simultaneously. You will need to determine the scope of the A/B test by indicating a starting and ending time. The recommended test range to get a representative sample and for the test results to be reliable is a minimum of two weeks. During the test period, pay attention to any potential external variants that may cause unusual interaction behaviors—such as weekends, a holiday, or any unexpected or catastrophic events in the area.

Once the testing period is over, you and your team will meet and discuss what to make of the results from the test. Because all the participating users are using the assigned interfaces at the same time (so, no seasonal factors here) and because they are randomly assigned an interface, you can have much more confidence that any difference in user behavior is indeed caused by the different content design. Statistical significance—i.e., a finding is *real* and not just happenstance due to luck—is the determination that a relationship between two or more variables is caused by something other than chance. Most testing management tools, like the ones we listed earlier, would do the stats for you and tell you if your results were statistically significant. Generally, you need to confirm that a difference has a p-value (probability value) of 5% or less to be considered significant. Should a comparison show statistical significance, you can add that finding to your overall content testing result to help with decision-making later.

What Is Heuristic Evaluation?

An expert review, or heuristic evaluation, is a quick and painless way to triangulate your test results with professional feedback. Unlike typical usability and A/B testing, the heuristic evaluation does not involve actual users but rather employs the expertise of those well versed in content design. It is a quick way to add a third perspective to your testing effort without the hassle of recruiting strangers and facilitating test sessions. Bear in mind, however, that this method lacks statistical significance, just like a traditional usability test, because it is not meant for generalization. Yet, heuristic evaluation can yield important feedback about a design from experienced evaluators who know what to look for and focus on.

The first step to conducting a heuristic evaluation is identifying your expert panel. It is best to gather feedback from those not already on your design team who know about the objectives and constraints of your project. This can help reduce bias in comments. Try seeking assistance from peers working in similar industries or professions, such as another UX writing agency or content design service provider, to learn about their perspectives. One strategy for doing this is by joining communities on social media and creating local professional networks (via industry networking events, conferences, etc.) to make long-term connections.

Once you have located your expert panel, begin the heuristic evaluation by giving them information about the scope of your project and, if necessary, a consent for non-disclosure agreement. Most designers and writers know about heuristic evaluation, but if you're working with someone who doesn't, provide them with the following guidelines built from Jakob Nielsen's early research (Molich & Nielsen, 1990; Nielsen, 1994).

1 **Visibility of system status:** The design should always keep users informed about what is going on through appropriate feedback within a reasonable amount of time.
2 **Match between the system and the real world:** The design should speak the users' language. Use words, phrases, and concepts familiar to the user rather than internal

jargon. Follow real-world conventions, making information appear in a natural and logical order.

3 **User control and freedom:** Users often perform actions by mistake. They need a clearly marked "emergency exit" to leave the unwanted action without having to go through an extended process.

4 **Consistency and standards:** Users should not have to wonder whether different words, situations, or actions mean the same thing. Follow platforms and industry conventions.

5 **Error prevention:** Good error messages are important, but the best designs carefully prevent problems from occurring in the first place. Either eliminate error-prone conditions or check for them and present users with a confirmation option before they commit to the action.

6 **Recognition rather than recall:** Minimize the user's memory load by making elements, actions, and options visible. The user should not have to remember information from one part of the interface to another. Information required to use the design (e.g., field labels or menu items) should be visible or easily retrievable when needed.

7 **Flexibility and efficiency of use:** Shortcuts—hidden from novice users—may speed up the interaction for the expert user, such that the design can cater to both inexperienced and experienced users. Allow users to tailor frequent actions.

8 **Aesthetic and minimalist design:** Interfaces should not contain irrelevant or rarely needed information. Every extra unit of information in an interface competes with the relevant units of information and diminishes their relative visibility.

9 **Help users recognize, diagnose, and recover from errors:** Error messages should be expressed in plain language (no error codes), precisely indicate the problem, and constructively suggest a solution.

10 **Help and documentation:** It's best if the system doesn't need any additional explanation. However, it may be necessary to provide documentation to help users understand how to complete their tasks.

Ask your expert panel to submit their evaluation either on an open-ended form or on a rated scale (or both) to you, and you may then compile these evaluations to create a synthesis of findings that indicate any changes necessary for your design.

What Is Validation?

Unlike previous usability testing, A/B testing, and heuristic evaluation methods, validation is usually deemed the final test to validate the completeness and functionality of a product. Think of it as the final engine check. The purpose of validation is to ensure that a given design is actually going to work in mundane, everyday situations, unlike those in controlled environments like usability testing and even A/B testing. Especially for products used in higher-stakes environments (such as medical technology), validation aims to ensure that all use-related hazards have been mitigated before being deployed in the real world.

To conduct validation, you should first identify the actual environment in which your content will likely be consumed or applied. Then, recruit a sizable number of the most representative users to conduct this final test. For medical and health-related content, the US Food and Drug Administration requires at least 25 participants who express a satisfactory experience in a validation test. Other projects can use this number as a benchmark.

You should strive to give test users the actual tool or hardware for accessing the content, instructions for use, and training materials (if applicable) to fully experience the content's design. Last, you need to ensure that the tasks performed by test users in the validation process include frequent/primary operational functions as well as critical/extreme scenarios. This will allow you and your client to validate the applicability of designed content across a range of uses.

What's Next? Strategies for Managing Content

Upon confirming that your content is capable of achieving the desired objectives, the next phase in UX writing is the deployment of the content. "Shipping" content into the real world lets you see how it will perform in the wild and allows your client to use the content for commercial or other purposes. In this launching phase, a key consideration is the management of content. As we've introduced in the opening chapter, content strategy, as part of the ecosystem of UX writing, is closely tied to content management. For some, like Mike Atherton and Carrie Hane (2018), content management is the end of theory and the beginning of decision and practice. Content management is defined as the process of choosing, applying, and leveraging appropriate systems for developing, containing, and publishing content.

Once your polished content is ready to meet the world, you need a *workhouse* (space) and a *workflow* (process) to push it into the spaces where your content will manifest and do what it's made to do. We begin here with the process, the workflow, for managing content. In a networked and collaborative age, we can no longer rely on siloed content development and storage. Consider the following scenario for context:

> Jifang, Noam, and Laura are writers at Lubbock International Bank, a growing financial institution designing new user-facing materials for their maturing customer base. Laura has been with the bank for more than 15 years. She is the communications manager who oversees external relations. Jifang and Noam were hired as social media specialists who specialize respectively in Finbook and Finster content. The team of three is the mouthpiece for Lubbock International, so to speak. Recently, there is a new initiative to expand Lubbock International service locations to include a small province in China, Hainan. Jifang is particularly excited about the prospect of translating existing and new content from English for the Chinese market.
>
> In the past, Laura the manager has asked Jifang and Noam to create the necessary content independently and publish the approved content separately at the appropriate time. With the new Chinese market, however, Jifang noticed that the translated content often contains mistakes and needs corrections. She is fine with helping the translators with editing for Finbook but Noam—who doesn't know Chinese Mandarin—has expressed frustrations with the need to correct his Finster content every time an updated set of translations is made. He would like to just draw from a shared, centralized "storage" where the most up-to-date versions of the content are available for him to publish at the scheduled time. Laura is a bit confused about Noam's frustration—why can't he just ask Jifang for the most recent version of the content?

Scenarios such as this are telling for many small-to-large companies that work with constantly renewing content. Moreover, these scenarios typically happen within companies that still employ individually created or edited content rather than shared, single-sourced content within their entity. What these companies could benefit from is a systematic approach to content management, which includes the following strategies:

- Single-sourcing
- Topic-based writing
- Using markup standards for dynamic content

All of these methods fall under the purview of structured authoring. As a UX writer, you will need a fundamental understanding of structured authoring to help your organization lessen unnecessary overlaps in workload and optimize content performance.

What Is Structured Authoring?

Structured authoring was made popular by Robert Horn, who introduced a "mapping" approach to writing information. In his earliest publications (Horn, 1966, 1989), he determined six (now seven) types of information:

- Procedure
- Process
- Concept
- Structure
- Classification
- Principle
- Fact

These information types are loosely related to the three core types—concept, task, and reference—developed by the technical publications department at IBM in the early 2000s. We will return to this specific standard in a little bit. Horn's information mapping practice inspired technical writers, particularly those who create documentation content for both end users and developers (including software designers, programmers, and other writers), to streamline their workflow using standardized conventions.

The idea is to apply a uniform way of organizing and storing content by marking it up with a hypertext scripting language, i.e., XML (extensible markup language), which is seen as a shared vocabulary among those who follow the same standard. Because the method is standardized and its rules are collectively followed, computers and applications can recognize the shared language. Therefore, content can be easily published in many formats for different devices, easily searched and retrieved from a common database, and easily edited or redesigned based on stylistic needs. Figure 7.5 shows a simple example of a marked-up food menu to indicate the type of information within the content.

Notice how the breakfast menu contains repeated tags to indicate categories of information—food name, price, description, and descriptions. This mapping convention allows for a systematic reading of the content by machines as well as humans (although it helps if you already know a bit of HTML scripting convention).

Another common example you may use as a mental reference is the English dictionary. Most publishers follow a conventional taxonomy (hierarchy) when putting together the thousands of entries in their dictionaries:

Term (the word/entry)
Pronunciation (the phonetics)

```
<breakfast_menu>
  <food>
    <name>Belgian Waffles</name>
    <price>$5.95</price>
    <description>Two of our famous Belgian Waffles with plenty of real maple
    syrup</description>
    <calories>650</calories>
  </food>
  <food>
    <name>Strawberry Belgian Waffles</name>
    <price>$7.95</price>
    <description>Light Belgian waffles covered with strawberries and whipped
    cream</description>
    <calories>900</calories>
  </food>
  <food>
    <name>Berry-Berry Belgian Waffles</name>
    <price>$8.95</price>
    <description>Light Belgian waffles covered with an assortment of fresh berries and
    whipped cream</description>
    <calories>900</calories>
  </food>
</breakfast_menu>
```

Figure 7.5 An example of how markup language indicates the type of information contained in the content, such as food, name, price, description, and the amount of calories.

Parts of speech (e.g., noun, verb, adjective)
Definition (the descriptive meaning of the term)
Example (an application of the term in a sentence or question)

Like the breakfast menu, this dictionary taxonomy helps users understand the content quickly. A standard structure makes finding and learning the information in the content easier.

Structured authoring champions topic-based writing. With information mapping, content writers and designers developed an approach to composing information—that is, chunking content into modular forms. One analogy for understanding this approach is to think about individual modules as LEGO pieces. Each topical content is akin to a single brick that can be combined with other pieces to build a larger structure. Similarly, topic-based content is a standalone, discrete entry composed of the following characteristics so that it can be plugged into a larger document or removed easily:

- focuses on one subject
- has an identifiable purpose
- does not require external context to understand

A widely adopted topic-based writing standard is the Darwin Information Typing Architecture (DITA) invented by the aforementioned technical writers at IBM and later donated to OASIS, an international consortium of vendors and users devoted to developing guidelines for interoperability among products that support the standard generalized markup language (SGML). Indeed, there are many acronyms floating around, but bear with us as we unpack DITA briefly. This markup standard may prove to be commonplace among your organization and team in your UX writing career.

DITA provides a schema for topic-based writing by defining the core categories of task, concept, and reference. A *task* provides detailed steps for users to perform that task and would include pre- and post-requisites, some context about that goal or task, choices, and/ or results. A *concept* explains how things work or what things mean. Concepts give users a way to understand how things fit together. Concepts often include graphics that explain or illustrate them, such as an architectural diagram with high-level information. *References* are quick lookups of information with lots of detail. These might include very technical graphics, organized lists (often alphabetical), and/or very detailed tables. Reference topics are meant for scanning, not reading. An example of a reference is a specifications table or a command and command options list (Samuels, 2013; Figure 7.6).

Apart from sticking to the standards when creating topics, the most important mindset to keep is to treat content as separate from its presentation (form). Many writers, including those trained in UX, tend to see writing as a form of display—the font style, font size, text alignment, textual treatment (underline, bold, italics), and other typographic elements. Structured authoring requires you to temporarily set aside these visual considerations and focus purely on the core text and its structure (organization) within the content. Other media, like diagrams and videos, are also treated as texts with extratextual information. Thinking in this minimalist way can help you focus on the substance and meaning of the text without the distraction of presentational needs. Once your core text is developed, you can then shift your attention to visual display by setting "rules" for appearance, such as:

Title: Arial, bold, 18pt. Followed by a line break/separator.
Heading 1: Arial, bold, 14pt.
Heading 2: Arial, italic, 14pt.
Paragraph: Times New Roman, 12pt.

These rules can be assigned to style guides using another markup language known as CSS (cascading style sheet). Similar to HTML and XML scripting languages, CSS uses tags to assign attributes and values to the marked content based on your predetermined style rules.

With information mapping and topic-based writing, UX writers can now practice single-sourcing as an efficient way to create, store, and deploy content. As Figure 7.7 shows, different modules—units of content like little LEGO pieces—make up a document, which can be published in multiple formats (print booklet, PDF, ePub, audiobook, webpage/ HTML, etc.).

Recalling the scenario where Jifang, Noam, and Laura were faced with challenges in updating content (in that case, translating content) in a seamless manner, they could benefit from a single-sourced writing approach. For instance, Laura could have imported all translated content to a shared site, where Jifang could make "master" edits if she found errors in the translations. Then, both Jifang and Noam could pull their respective Finbook and Finster content from the shared site (a *single* source). This would eliminate the need for Noam to ask for the corrected content from Jifang separately, thus reducing redundancy, frustration, and errors due to version control. For larger organizations, single-sourcing saves not only time but, more importantly, money and avoids the above-stated problems in content delivery.

To adopt a structured authoring workflow, you need an appropriate CMS for managing your content. In the next section, we show the different types of CMS so you can make informed CMS selection decisions for your team.

DTD input

```
<?xml version="1.0" encoding="UTF-8"?>
<!DOCTYPE task PUBLIC "-//OASIS//DTD DITA Task//EN""task.dtd">
<task id=t-marinara">
        <title>Marinara Sauce</title>
        <shortdesc>Prepare a crowd-pleasing red sauce for pasta in about 30 minutes.</shortdesc>
        <prolog>
                <author>Carlos Evia</author>
                <metadata>
                        <category>Italian</category>
                </metadata>
        </prolog>
        <taskbody>
                <prereq>
                        <ul>
                                <li>2 tbsp. of olive oil</li>
                                <li>2 cloves of garlic, minced</li>
                                <li>1/2 tsp. of hot red pepper</li>
                                <li>28 oz. of canned tomatoes, preferably San Marzano</li>
                        </ul>
                </prereq>
                <steps>
                        <step>
                                <cmd>Heat olive oil in a large saucepan on medium.</cmd>
                        </step>
                        <step>
                                <cmd>Add garlic and hot red pepper and sweat until fragrant.</cmd>
                        </step>
                        <step>
                                <cmd>Add tomatoes, breaking up into smaller pieces.</cmd>
                        </step>
                        <step>
                                <cmd>Simmer on medium-low heat for at least 20 minutes.</cmd>
                        <step>
                                <cmd>Add parsley.</cmd>
                        </step>
                        <step>
                                <cmd>Simmer for another five minutes.</cmd>
                        </step>
                        <step>
                                <cmd>Serve over long pasta.</cmd>
                        </step>
                </steps>
        </taskbody>
        </task>
```

PDF output

Marinara Sauce

Prepare a crowd-pleasing red sauce for pasta in about 30 minutes.

Ingredients
- 2 tbsp of olive oil
- 2 cloves of garlic, minced
- 1/2 tsp of hot red pepper
- 28 oz of canned tomatoes, preferably San Marzano
- 2 tbsp of parsley, chopped

Steps
1. Heat olive oil in a large saucepan on medium.
2. Add garlic and hot red pepper and sweat until fragrant.
3. Add tomatoes, breaking up into smaller pieces.
4. Simmer on medium-low heat for at least 20 minutes.
5. Add parsley.
6. Simmer for another 5 minutes.
7. Serve over long pasta.

HTML output (with CSS)

MARINARA SAUCE

Prepare a crowd-pleasing red sauce for pasta in about 30 minutes.

Ingredients
- 2 tbsp of olive oil
- 2 cloves of garlic, minced
- 1/2 tsp of hot red pepper
- 28 oz of canned tomatoes, preferably San Marzano
- 2 tbsp of parsley, chopped

Steps
1. Heat olive oil in a large saucepan on medium.
2. Add garlic and hot red pepper and sweat until fragrant.
3. Add tomatoes, breaking up into smaller pieces.
4. Simmer on medium-low heat for at least 20 minutes.
5. Add parsley.
6. Simmer for another 5 minutes.
7. Serve over long pasta.

Figure 7.6 Marinara sauce DITA task (top) transformed into a PDF deliverable (bottom left) vs. an HTML deliverable (bottom right). The HTML output includes a link to an external CSS file that indicates the formatting rules—like font style, alignment, and text justification—for the output (Evia, 2019).

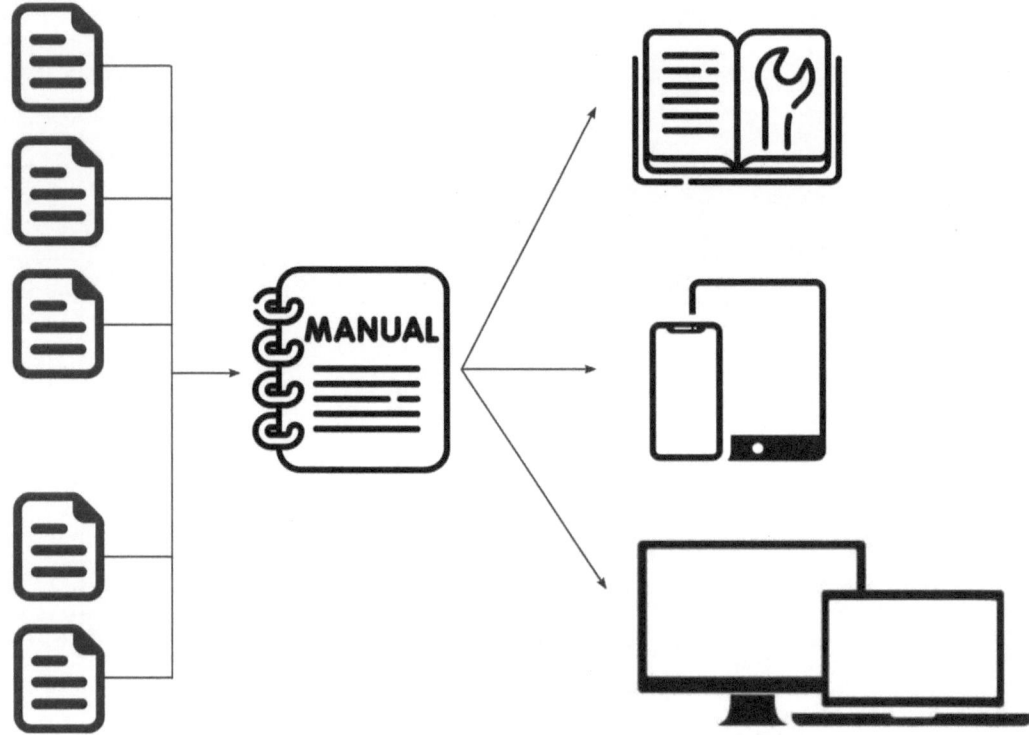

Figure 7.7 Single-sourcing workflow enabled by structured and topic-based writing.

What Are Content Management Systems?

CMSs are the *workhouses* designed to help create order and manage collaboration in your content deployment process. Most of these software applications are web/cloud-based and do not require downloading to your machine. There are three main types of CMS:

- Component CMS
- Document management system
- Web CMS

As Figure 7.8 shows, these three types of CMS work in tandem with one another. Component CMSs are applications that manage the source materials of your content. They offer a centralized online location where your sources are housed and can be directly edited and saved to reduce version control troubles. This kind of system is typically used at the enterprise level and requires business subscriptions. They usually come with premium support that helps companies ensure compliance with content requirements (especially in medical, tech, and safety products), content auditing, and optimization for monetization purposes. Popular components of content management platforms include:

- Xyleme (https://xyleme.com/)
- Paligo (https://paligo.net/)
- XML Documentation for Adobe Experience Manager (https://www.adobe.com/products/xml-documentation-for-experience-manager.html)

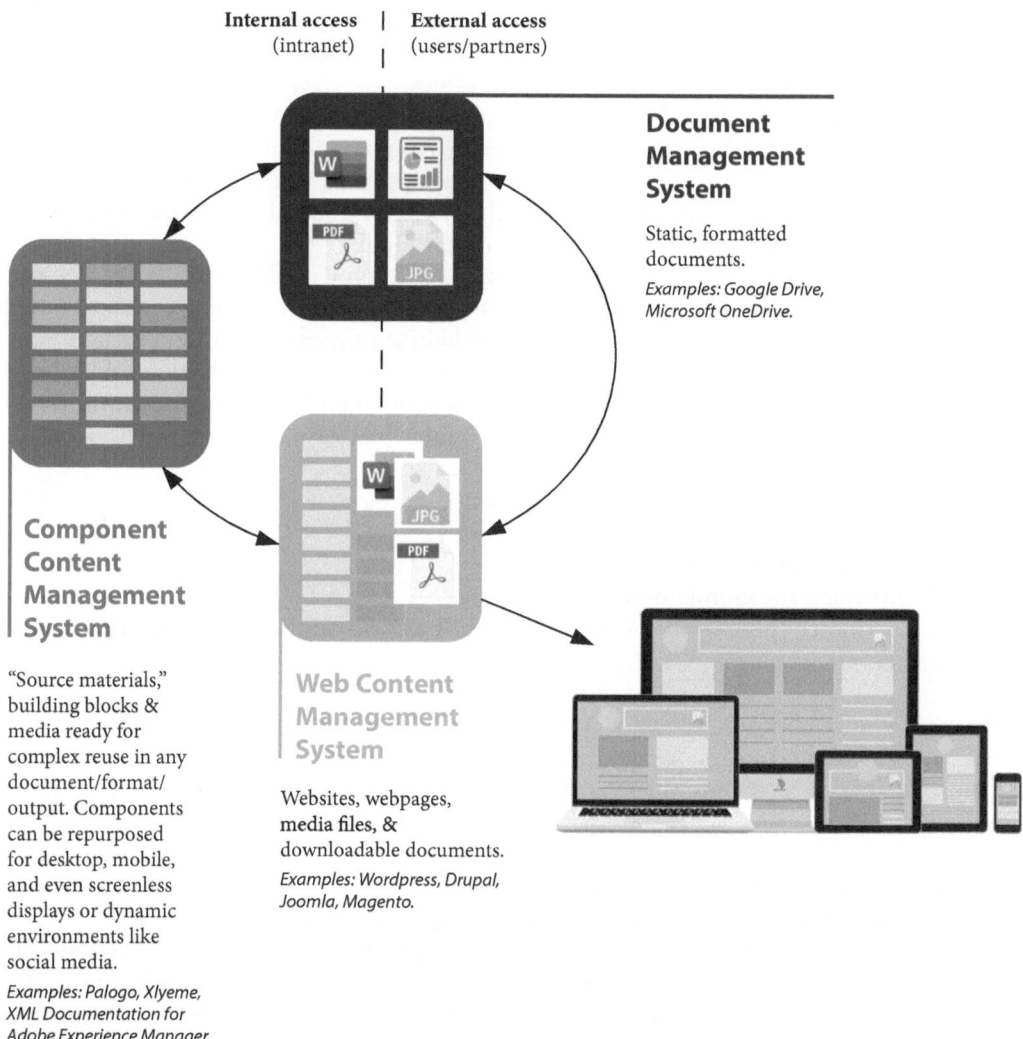

Internal access | **External access**
(intranet) | (users/partners)

Document Management System

Static, formatted documents.
Examples: Google Drive, Microsoft OneDrive.

Component Content Management System

"Source materials," building blocks & media ready for complex reuse in any document/format/ output. Components can be repurposed for desktop, mobile, and even screenless displays or dynamic environments like social media.
Examples: Palogo, XIyeme, XML Documentation for Adobe Experience Manager.

Web Content Management System

Websites, webpages, media files, & downloadable documents.
Examples: Wordpress, Drupal, Joomla, Magento.

Figure 7.8 Three main types of CMS.
Source: Adapted from Tham, Howard, & Verhulsdonck (2022).

If you're a student, you should also have experience with another kind of CMS that is essentially a component CMS, i.e., the learning management system (LMS), like Canvas, Blackboard, Moodle, and D2L Brightspace. One distinctive difference in LMS is the affordability of instructor grading. Both the instructor and students interact with learning materials in an LMS, just as designers and users interact with content in a component CMS.

The second type of CMS is a document management system, which is used to organize digital assets. Unlike component content management, document management focuses on rich media storage, editing, scaling, and delivery. Digital assets include textual content (like word documents, PDFs, transcripts, alt-texts, and other microcontent), audio files (music, jingles, and alert sounds), image files (pictures, charts, and GIFs), artwork (design files, vector graphics, and icons), presentations (slide sets, static, and animated slides), video files (movies and clips), and data files (spreadsheets, tables, and raw/source files). A document management system contains robust metadata forms and filters to help with systematizing documents and enabling easy search and retrieval. An example of an in-house document management system is Elevator (http://www.elevatorapp.net/) at the University of Minnesota, which features flexible format support. Other popular document management systems include cloud drives like Google Drive and Microsoft OneDrive.

A web CMS is perhaps the most popular and essential type of CMS that many UX writers are already familiar with. The main function of a web CMS is to manage website content creation, maintenance, and removal. It serves the hybrid purpose of component and document management by providing backend spaces where users can tag, organize, and design the architecture of a content site. In Chapter 10, we include recommendations on using a web CMS to create your UX writing portfolio. Popular platforms include:

- WordPress - web-based (https://wordpress.com/)
- WordPress - server-based (https://wordpress.org/)
- Drupal https://www.drupal.org/
- Joomla (https://www.joomla.org/)
- Adobe Commerce (previously Magento) (https://business.adobe.com/products/magento/magento-commerce.html)

Whether you're choosing to use a component CMS, document management system, or web CMS, you should give consideration to roles and access. CMSs allow you to assign specific user roles that dictate their levels of access to content within the CMS. Common roles include:

- Owner: Responsible for creating and editing content.
- Editor: Responsible for tuning the content message and the style of delivery
- Administrator: Responsible for managing access permissions to folders, collections, and files, usually accomplished by assigning access rights to user groups or roles.
- General user: Reads or otherwise consumes the content after it is published or shared.

Deploying Omnichannel Content

So now that you are familiar with how you need to test your content through usability testing and know that a content strategy requires you to work with CMSs *workhouse* (space) and to develop a *workflow* (process) to push out content, it is important to also consider *where* that content can appear. As you learned earlier about *microcopy* (content that appears in an interface) and *microcontent* (content that can quickly be consumed in a few seconds and help solve a problem a user is having), it is important to consider how your content is deployed. Specifically, you want to differentiate between the multiple channels through which your users can access your content.

Think about all the channels that you use in your everyday life to communicate and get information:

- Social media
- Websites

- Email
- SMS
- Work conversational channels (Slack, MS Teams)
- Phone Calls
- Apps
- Digital voice assistants, such as Alexa
- Virtual or augmented environments (VR/AR)

All these channels have different characteristics that make them suitable for particular types of content. If you want to contact your friend quickly, you will probably send them an SMS. If you have a substantial message you need to share with your boss that is important to read in detail, you'd rather send an email than an SMS. As media theorist Marshall McLuhan (1994) noted, "the medium is the message"—the characteristics of a medium will shape how messages appear. Hence, how we perceive a medium and its characteristics will impact how we communicate. The same principle goes for content—different media channels will provide us with different ways to push out content. We can send out short messages via SMS or social media, whereas longer messages can be sent via email or websites.

Now to add one more consideration: with more medium channels to communicate in our everyday life clamoring for our attention, our expectations of increasingly personalized communication have also increased. Simply put, we don't have limitless attention and expect content to be tailored to a medium and to be effective in quickly letting us get on with our day with minimum interruption. In other words, we want *content that works for us personally*. That is, we want content that is *hyper-personalized*.

1 We want to be identified instantly on our preferred device.
2 Have immediate access to our information history whichever channel we decide to access that information (web, app, mobile device, and phone).
3 Provide intelligent content that solves our problems quickly with appropriate context for what we are trying to get done.

(Iero, 2022)

As you can tell, there are a lot of expectations for deploying content. To help you understand this, we introduce the UX writing-attracting loop (Figure 7.9).

The task of UX writers is to manage those multiple channels so content is deployed in an intelligent way to the user at the right time, and the right moment. With the triple revolution of wireless networks, ubiquitous internet, and mobile devices making communication a thing at our fingertips at any moment, we also want intelligent content particular to what we are doing at that moment. Indeed, some researchers hypothesize we are in the age of "networked individualism"—where users relate to the world primarily through a networked sense with information at their fingertips delivered at incredible speeds (Rainie & Wellman, 2012). Simply put, users aren't expecting to just get their information from your website. They want it on their phone, through an app or SMS message, at their convenience. When they place an order with UberEats, they want up-to-date information when their delivery is there on their preferred method of communication. When they order a package, they want to know where it is and when it is arriving. All this is part of a content strategy that uses location information, messages, and microcontent.

Hence, UX writers know that users want content personalized for them in near real-time. What they also know is that users want to get content on their preferred channel. For that,

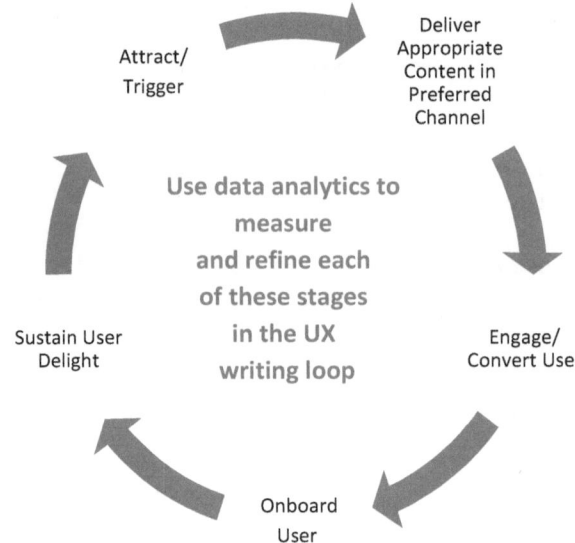

Figure 7.9 The UX writing-attracting loop triggers the interest of the user, delivers appropriate content to the user in their preferred channel, engages that user, onboards them successfully to their channels, and sustains their interest using data analytics.

UX writers need to consider an *omnichannel strategy* that manages to deliver accurate, personalized information to the user in near-real-time and in their preferred channel of communication.

Real World Snapshot 7.2: The Case of AR-Bike and Omnichannel UX

Consider this scenario: Juan is a UX writer. He has been asked by his employer to develop some content for his company, AR-Bike. The company's focus is on selling exercise bikes that feature a screen that displays an online instructor that motivates them and visuals of different biking terrains. AR-Bike wants to roll out a new feature—one-on-one mentoring—in their exercise software and wants people to be excited about trying this new feature. Juan has been asked to develop content that should be dispersed across multiple channels, such as their website, their social media account, be accessible to users on their app, and be available on the company's own website. He also knows that users sometimes have specific questions about their bike that are more technical and that they prefer to use multiple channels to contact the company. Some people prefer asking a question via the app, others call up; and yet others like sending a text message for an answer. What's more, a lot of these questions may require follow up by AR-Bike in an integrated manner. What starts as a message in an app may lead them to a conversation via chat with an actual representative on the phone, who may need a technical lead to answer the question. If the customer has to repeat what they've already stated in the app again in the phone conversation, they feel like they are not being heard. What is needed is an omnichannel strategy to make sure the user feels like the interaction is seamless from one medium to the next and personalized specifically to them as a user.

In other words, Juan needs to come up with a plan to manage the multiple channels and turn them into an omnichannel experience for his users. To make it an omnichannel experience, he needs to manage these multiple channels as if they were one experience for the user through his CMSs while keeping track of user interaction data. For this reason, many companies are investing in customer relationship management (CRM) systems such as Salesforce, Salesflare, Zoho, or Hubspot. The value of these CRMs is that they keep track of, and make transparent, user events as histories. Such CRMs also have implications for UX writers, who have to consider writing reusable content across different channels while keeping in mind user histories.

You may wonder what the difference is between multichannel and omnichannel:

- **Multichannel:** Multiple channels are being used to address the user, e.g., web, app, phone, chatbots, and text messaging, but they are not used in an integrated manner to understand the user and their journey. This sometimes leads to frustration for the user, who may be forced to restate a problem they've already reported to the company in a different medium earlier.

- **Omnichannel:** Multiple channels are utilized in an integrated manner that understands the user journey in a seamless manner, so the interaction feels personalized to the user. Through the use of CRM systems, the user's personal journey, history, and context of the interaction are shared between channels, and the user is directed to the appropriate person specialized in their particular issue with access to the user journey (Figure 7.10).

Because companies know that users have more channels to choose from to contact them and want to make it easy for them to engage and feel the interaction is personalized to them with the company no matter what medium, they use omnichannel strategies. Some people are video or phone shy and would rather send a message through the app, or they are okay with sending a text message. The point of the omnichannel strategy here is to

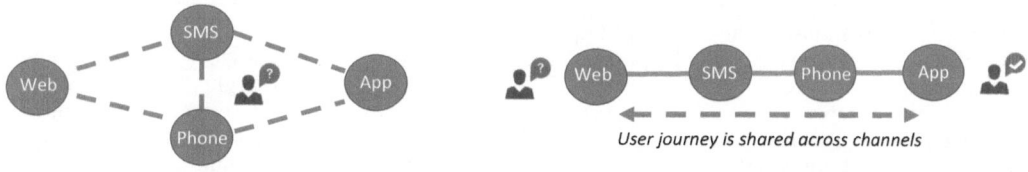

Multi-channel approach *has channels operate independently.* There is no understanding of the user journey, causing friction because the user has to restate their issue each time to get it resolved when they switch channels.

User journey is shared across channels

Omnichannel approach *is where channels are connected and aware of the user journey.* Because the user's issue is known to the system, the experience feels personalized to the user and each channel can give contextualized help knowing the user's history across channels.

Figure 7.10 The difference between a Multichannel approach vs. Omnichannel approach is that the latter keeps a history of the user's interactions accessible across all channels.

meet these people at their preferred channels and provide them with a seamless experience no matter which medium they choose. Companies will want to avoid, in other words, a situation where each channel is siloed and one side of the company doesn't know what the other side of the company is saying to the customer. A lot of omnichannel strategies are supported by artificial intelligence, data analytics, and increased storage of user interactions to create a profile history of the user. For example, omnichannel software such as Twilio, SendGrid, and Segment let you interact with users from a single platform that facilitates email, SMS, video, chat, voice, and other mediums while also storing past interactions.

Omnichannel content is hard to do, but easier once you start thinking in terms of how a CMS allows for transferring the same content into a website, PDF, text message, or app interface. When this capability to reuse content is then paired with CRMs that offer specific context on the user's journey and can match specific content to help that user, you are seeing how data analytics, artificial intelligence, tracking the user, and matching that with tailored content can help improve the UX. The omnichannel strategy, then, is to add real-time information that users need and to do so in an integrated and seamless manner so the experience is pleasant for the user.

Walgreens' Omnichannel: A good example of an omnichannel strategy is that of the popular United States-based pharmacy chain Walgreens. To refill prescriptions, customers can call their local Walgreens and provide their prescription information by phone. However, they can also do this through an app by scanning the barcode on their bottle. Alternatively, if they are about to run out of their prescription, Walgreens asks them if they want to refill the prescription in an SMS. The user just needs to type in "Yes," and Walgreens automatically contacts their doctor to ask for refill permission, if needed. In case the user has questions about the medication, they can call Walgreens' personnel, who will have their information on their screen as soon as they call and their phone number is recognized. The point of such an omnichannel strategy is that the user can decide which channel they prefer, and their experience is always supported by real-time information the company has access to so as to hyper-personalize it for them (Table 7.2).

Table 7.2 Pharmacy chain Walgreens' Omnichannel to provide hyper-personalized omnichannel content for helping customers easily refill their prescriptions through their preferred channel

Omnichannel strategy	SMS	Phone	Smart scan	App	Website
Scenario: to refill a prescription at Walgreens Pharmacy	Respond to the automated SMS to reply with "REFILL" if you want to refill your prescription	Call and inquire by phone menu	Scan code of prescription through Walgreens App to automatically refill	Send a message through Walgreens App that you want to refill	Visit the Walgreens Website and log in to your personal account
UX writing content required	SMS confirmation message	Chatbot menu dialogue tree by phone; conversational text	Confirmation message via App	Confirmation message via App	Confirmation message via website

Obviously, this requires UX writers to write content that can work across these channels and to employ intelligent content responsive to the user's needs that takes into account their personal history (because it has to store the refill of their prescriptions and know which prescriptions work and do not work in combination).

Real World Snapshot 7.3: Omnichannel Approaches for Food Delivery Service, Deliveroo

Deliveroo Omnichannel: Deliveroo, a food-based delivery service (similar to UberEats) that operates in Europe, Australia, Asia, and the Middle East, also uses an omnichannel strategy. Because Deliveroo delivers food to a specific address, all sorts of things can go wrong, and they want to avoid the "hangry" customer. If someone is hungry and angry, that is definitely not good. So Deliveroo offers multiple touchpoints for users to inquire about and communicate with the company as part of their omnichannel strategies. For example, the delivery person could have trouble finding the address because the restaurant sent the wrong food order and the person who placed the order wants to track where their order is because they are hungry or have plans for the night and want to find out when they can get their delivery. A lot of this requires an omnichannel strategy to help people get in touch with their deliverer to avoid any frustration and to personalize information between the user, the deliverer, and the company. Hence, Deliveroo uses an omnichannel strategy (see Figure 7.11, Table 7.3).

Deliveroo Omnichannel Strategy for Food Deliveries

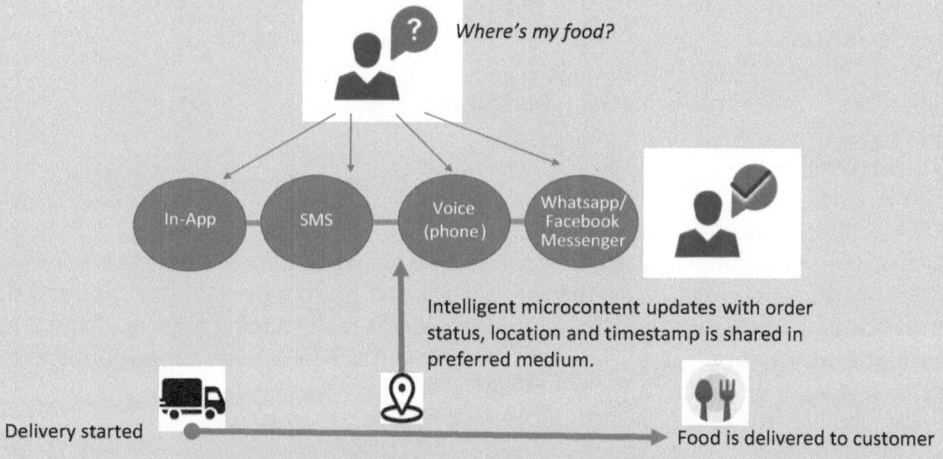

Figure 7.11 Deliveroo uses omnichannel strategies so the user, no matter their preferred channel, always gets contextualized information referring to their user journey (when their food will arrive).

Table 7.3 An omnichannel strategy lets customers pick their preferred channel for communicating and is always informed with real-time information to help customers in contextualized ways at multiple touchpoints in their user journey

In-App messaging	SMS	Voice	Whatsapp/Facebook Messenger
If users use the Deliveroo mobile app for their order, they get push notifications on their mobile device on their order status and whether the deliverer picked up their order, is in transit to the customer, or delivered it.	Deliveroo sends out promotions, order updates, and service updates via SMS. These updates use customer data, so promotions are about products they care about, which may prompt them to order.	Customers can call their deliverer if they have a problem with the order or want to ask a question. Phone numbers are masked and private, but allow the deliverer and user to communicate directly on the order.	Users can order via the website and decide to opt-in to get notifications via social media messaging systems (Facebook Messenger and Whatsapp).

Source: Test (2021).

Obviously, deploying content is a lot more complex with an omnichannel strategy. However, with the UX writing loop, we hope to give you a sense of how this works. There are a great many channels that you can write content for, such as:

- Print
- Websites
- Mobile devices
- Apps
- Digital Voice Assistants/Conversational agents
- Chatbots
- Touchscreens
- AR/VR

To give you an example, chatbots require you to write content that is conversational (non-technical), contextual (meets the user at the exact point where they are at), and derives an understanding of the specific information the user is really asking for (e.g., "can you look up my flight information") and that can be deployed across a website, a mobile handheld device interface, or as part of a phone menu.

To Push or to Pull, that Is the Question

Next to knowing how to create a strategy to deploy across multiple channels, as part of omnichannel, you will need to deploy smart strategies to reach users. Ideally, you will develop intelligent content—content that meets the user where they are at that exact moment. There are two strategies important in meeting the user, and they are push and pull strategies. These strategies are embraced by marketing, but you will also find them useful for UX writing as you think about how you are meeting with your user.

Like in everyday life, when you meet someone new, it's important to make a good first impression.

- A push strategy means you push a message out to the user. Often, a *push* strategy is meant to foster awareness. The advantage of this is that you have control over the messaging and can do this to reach a great deal of people. The disadvantage is that people might feel like you are invading their privacy or haranguing them with your message.
- A pull strategy means you pull the user to your message. Often a pull strategy is meant to engage the user with the brand in an ongoing manner. The advantage of this is that your users are coming to you. The disadvantage is that you had better have excellent content to sustain their interest.
- A viral strategy means a strategy intended to be shared between different social media networks. Often, the viral strategy requires a lot of insight into one's audience. The advantage of this is that your users push out your content to their own social media connections. The disadvantage is that you do not have a good view of who you are specifically targeting.

As an example, many companies use push-and-pull strategies together. First, a push strategy is deployed by creating advertising space on a social network. Often, the message is placed in the eyesight of the user based on their searches. For instance, you may have been looking for that perfect pair of shoes from your favorite brand. A push strategy could mean that you would see that image in your social media feed as a result of an e-commerce company that sells shoes. Once you click on their link, the pull strategy of the company can mean that you see images of first-time visitors who get a 10% deduction on their purchase or that becoming a member saves you 25% on each purchase (a good deal, if you ask us).

Alternatively, some clever UX writers create strategies that go viral, where the pushing is done by the user. For example, the company Wendy's is well-known for its social media presence on Twitter. By roasting people, Wendy's gets free advertising as people comment upon it (pull strategy), then share their social media post (push strategy), and in turn, someone else shares it with someone else (viral strategy).

Likewise, as a UX writer, you can think about how push-and-pull strategies can work together to create content that helps the user and gets them into your ecology. For example, you can use quick pop-up messages to inform your user that something needs their attention (a push strategy), and following up, you can ask them if they want to subscribe to your newsletter (a pull strategy). The next chapter will describe how you can measure whether your content is doing well and how to get reusable users to keep returning.

Conclusion

In this chapter, you learned about three crucial phases for a UX writer: (1) to test (or measure), (2) to manage, and (3) to deploy content that works for the user. As part of the first phase of testing content, you have learned about usability testing protocols, which can be in-person or remote. In usability testing, it is required that you have a script or protocol to conduct the test and get informed consent from your participants that they agree to be part of your usability test. As you learned, usability testing can be qualitative (featuring impressions of participants) or quantitative (involving numerical data on task performance or time-on-task). In the second phase of managing content, you learned about how the process of developing content takes place and which standards exist. You were introduced to a CMS and how this allows UX writers to follow standards while having a single repository to store their work and share it with others, and some standards exist for this (e.g., DITA, SGML). Lastly, in the third phase, you also learned about how you can deploy content. Nowadays, this is part of an omnichannel strategy. You learned that users want content at the right time/place on a medium and device of their preference. This calls for an omnichannel strategy that is not only aware that multiple channels exist but also need to integrate them meaningfully for the user as one. Now that you have deployed content, it is important to be able to track what your users are doing. In the next chapter, you will therefore learn more about how to track your users using data analytics and some techniques to ensure they keep on returning to your content.

Chapter Checklist

This chapter has covered three phases for content, namely testing, managing, and deploying content.

- *Perform usability tests to find out more about what your user is thinking or doing.*
  - Make sure your participants sign the consent form, indicating that they agree to be a part of your usability test.
  - Use a scenario to set the scene for the participant and develop a script for them to follow.
- *Use qualitative measures to find out about your users' impressions.*
  - Use qualitative measures such as interviews to find out users' impressions of interacting with your design/content.
  - Use qualitative instruments such as SUS or TLX to measure what your users thought about your design/content in terms of usability or how difficult the task was for them.
- *Use quantitative measures to find out about your users and their behavior.*
  - Use time-on-task or clicks per task to measure how your user is performing a particular task.
  - The severity of the measure is important to note, as this indicates whether the issue is small (took longer time on task) or severe (could not complete a task).
- *Evaluate and test your content using different techniques.*
  - Use A/B testing to compare two different interfaces to see which performs better with users.
  - Use a heuristic (or expert) evaluation where your experts evaluate the content using the 10 usability criteria developed by Jakob Nielsen.
- *Develop content management system criteria for what constitutes quality content.*
  - Ensure that you and your colleagues use a single repository for content and follow different standards for it (e.g., DITA/SGML).
  - Use structured authoring principles and distinguish Horn's seven different types of information when developing your content.
- *Deploy your content while keeping in mind that your users are used to omnichannel approaches.*
  - Consider whether your users want to be instantly identified on their preferred device and channel of choice (e.g., mobile device, desktop, phone, etc.) and develop a plan to provide them with intelligent content that is personalized to the user in a systematic manner.

- Distinguish multi-channel approaches from integrated omnichannel approaches (which are harder to do).

- *Use the UX writing loop to ensure you attract and maintain the interest of your user by offering consistent and engaging content in various stages.*

  - You should create content that attracts, delivers appropriate content in the preferred channel, converts the user, onboards the user successfully to your platform, and sustains them with regular content that is useful or engaging to your user.
  - Use data analytics to measure, at each stage of the UX writing loop, how you can refine each stage.

Discussion Questions

1 Where should content usability testing take place? How do you decide?
2 Why is five the "magic number" for the number of participants for usability testing?
3 What are the different types of content management?
4 How does single-sourcing work? What are its benefits and drawbacks?
5 What does omnichannel content design mean for the user, and how is it different in approach from multi-channel content design?

Learning Activity

Find a website or app that you think needs work. Together with a friend, use heuristic evaluation (Jakob Nielsen's ten usability criteria) to evaluate its pain points and compare your notes. Then, together, address these pain points and plan a redesign. Write a script of what you would present differently to a user and what type of test you would give after having the user try out the new version.

1 What did you find out about the website /or app that needs improvement? Where did you converge or diverge in your ideas?
2 Where did you find elements that were most crucial to redesign to improve usability? Which elements were less crucial and more cosmetic?
3 What type of test would you apply after having your user go through your prototype? Why?

References

Atherton, M. & Hane, C. (2018). *Designing connected content: Plan and model digital products for today and tomorrow*. New Riders.

Evia, C. (2019). *Creating intelligent content with lightweight DITA*. Routledge.

Horn, R.E. (1966). A terminal behavior locator system. *Programmed Learning, 1,* 40–47.

Horn, R.E. (1989). *Mapping hypertext: Analysis, linkage, and display of knowledge for the next generation of on-line text and graphics*. The Lexington Institute (available from Information Mapping, Inc.).

Iero, T. (2022). What is hyper-personalization and why is it a key customer experience component? Retrieved from: https://getmindful.com/blog/hyper-personalization-key-customer-experience-component/

McLuhan, M. (1994). *Understanding media: The extensions of man*. MIT Press.

Molich, R., and Nielsen, J. (1990). Improving a human-computer dialogue, *Communications of the ACM, 33*(3) (March), 338–348.

Nielsen, J. (1989). Usability engineering at a discount. In G. Salvendy and M.J. Smith (Eds.), *Designing and using human-computer interfaces and knowledge based systems* (pp. 394–401). Elsevier

Science Publishers. https://media.nngroup.com/media/articles/attachments/Discount_Usability_paper_1989.pdf

Nielsen, J. (1994). Heuristic evaluation. In J. Nielsen and R.L. Mack, (Eds.), *Usability inspection methods*. John Wiley & Sons.

Nielsen, J. & Landauer, T. (1993). A mathematical model of the finding of usability problems. In B. Arnold, G. van der Veer, and T. White (Eds.), *Proceedings of ACM INTERCHI'93 conference*. Amsterdam, The Netherlands, 24–29 April 1993 (pp. 206–213). Association for Computing Machinery. https://doi.org/10.1145/169059.169166

Rainie, H., & Wellman, B. (2012). *Networked: The new social operating system*. MIT Press.

Samuels, J. (2013). Getting started with topic-based writing. Tech Whirl. https://techwhirl.com/getting-started-with-topic-based-writing/

Test, L. (2021). The ultimate guide to omnichannel. Twilio. Retrieved from: https://www.twilio.com/blog/new-guide-the-ultimate-guide-to-omnichannel

Tham, J., Howard, T., & Verhulsdonck, G. (2022). Extending design thinking, content strategy, and artificial intelligence into technical communication and user experience design problems: Further pedagogical implications. *Journal of Technical Writing and Communication, 52*(4), 428–459. https://doi.org/10.1177/00472816211072533

Usability.gov. (n.d.). Recruiting usability test participants. https://www.usability.gov/how-to-and-tools/methods/recruiting-usability-test-participants.html

Virzi, R.A. (1992). Redefining the test phase of usability evaluation: How many subjects is enough? *Human Factors, 34*, 457–468.

8 Tracking and Measuring Success

Chapter Overview

This chapter highlights ways to keep track of content and measure success. It introduces technical and analytical tools that help track content performance. It is important to develop content that attracts, engages, and sustains the user's interest. For this reason, UX writers need to know how to track their users and how to measure the success of their content. Tracking the user is done through various technologies, such as cookies, Urchin Tracking Modules (UTMs), and web analytics, to better understand your users and their behaviors. Measuring of content is done by common measurements (or metrics), called key-performance indicators (KPIs), that show whether users engage with your content and how your content performs. Together, tracking and measuring help to figure out what your users are doing, how your content performs, and what needs improving. This chapter presents these techniques so that you can better address users and their experiences and sustain content that engages them.

Learning Objectives

- Develop an understanding of a content creation framework to attract, engage, and sustain the user.
- Understand the difference between tracking and measuring.
- Distinguish different components for tracking and measuring content.
- Distinguish metrics from KPIs that help measure successful content versus indicating an organization's overall goals.
- Understand the 3×3 Method to organize goals, signals, and metrics and differentiate attitudes and behavior metrics for users from overall task metrics.

Attract, Engage, and Sustain Your User: Creating a Content Framework

Today, it is important to *attract*, *engage*, and *sustain* the user's attention. As you have learned in the previous chapter about the UX writing loop, it is crucial to keep improving your content. However, with social media, increased mobile device use, and limitless wireless connectivity, as well as various websites, apps, and interfaces to interact with, users also have many demands on their attention but little patience for poorly designed experiences. Users want to get moving with little interruption and an easy-to-understand process that

DOI: 10.4324/9781003274414-10

lets them do what they want to do. As a UX writer, you will need to carefully consider how this fickle audience expects you to attract, engage, and sustain their attention in a pleasant and efficient manner. For that, you need to develop a content creation framework that outlines how you are going to engage your users (Hubspot, 2022).

If you are working for a company as a UX writer, you will not only need to attract new users but also engage existing ones and sustain them with new content they will want to see. For example, you will want to create a social media presence and share posts with your users over a longer period of time so they feel you are worth their time. Nowadays, organizations know that audiences need to be engaged and have a lot of options, so UX writers need a content creation framework to plan out how they are going to strategically develop content over time and engage and sustain their audience. This ranges from writing content, planning a schedule, dividing up the work, reviewing the work, and keeping track of how the content performs. To do so, the content creation framework involves a number of different steps:

1 **Content:** What content needs to be developed for your particular audience.
2 **Planning a timeline:** When content needs to be created and distributed to your audience over time.
3 **Workflow:** Who handles content from initial creation to when it is published.
4 **Reviewing and editing:** Whether content aligns with your overall framework, style guide, and voice and tone.
5 **Deploying and tracking:** How content is performing by measuring how many people liked, commented, or shared it with others and tracking this performance on a continual basis.

The latter step is what this chapter will focus on so that you can deploy content, *track* how it is doing, and *measure* it on an ongoing basis so you can improve your content.

Tracking Your Users Online to Better Understand Their Attitudes and Behaviors

A key element of a UX writer's duties is to make sure the user experience is delightful and user-friendly. Part of this involves *tracking* how users interact with content. A key component of tracking involves technologies that track the user's personal information, their online behavior and attitudes, and how they got to your content. In the era of big data, companies such as Google and Meta/Facebook (among others) keep track of the personal data of people who use their platforms. This can include "housing type, age, sex, income, spending habits, hobbies and interests, items bought, items you're interested in buying, if you're traveling soon, and other data that may be fairly revealing" (Beck, 2015, p. 125). But tracking can also reveal what you are doing when you are on a website or app and where you are coming from in terms of sites and searches preceding your visit. As you can tell, tracking can tell a lot about your users and what they are doing!

The way tracking is done is by creating a personal profile of the user and their online actions through various technologies, such as:

• **Cookies:** When visiting a website, this stores a cookie on your hard drive that keeps track of your interactions on that website and others.
• **UTM:** When you visit a link from another site, UTMs stored in the URL of the website can keep track of which website you came from and which advertisement or website

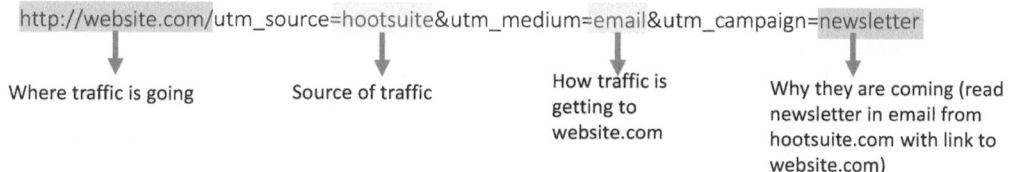

Figure 8.1 A Urchin Tracking Module (UTM) tracks where your site visitors are coming from and how they get to you.

Source: Emailonacid.

campaign resulted in traffic. Google Analytics uses this to keep track of where traffic is coming from (see Figure 8.1).

- **Website Statistics/Analytics:** Using UTMs and cookies gives you opportunities to use website statistics, such as how many visitors visited the site, what device they use, how long they were on your site (using cookies), and where they came from (using UTMs).

In other words, your "digital twin" exists out there that is a combination of data on your online *behavior* (what parts of the site you visited, what you bought), your *attitude* (what you like, what your interests are), and your *pathways* (how you got to the content through searches or other sites).

Data Analytics Help You Understand Users in Their Journey

As we noted in the earlier chapters, as a UX writer, you must understand and empathize with your audience. Using data and analytics is one way to better understand your audience. Below, we give some background on how data came to be so important to UX writing. A lot of the techniques to understand your audience were developed in digital marketing analytics, which seeks to understand what customers (users) are doing. As Kevin Hartman, Chief Analytics Evangelist and Head of Digital Marketing at Google, puts it, "data have become the most valuable asset for anyone who makes—or attempts to influence—a decision" (2020, p. 26). For some, data have become more valuable than oil because it helps better understand your audience and more effectively connect with them, which many businesses want. The same goes for UX writers, who want to ensure they capture their audience's attention and understand what they need.

Although digital marketing often has a "bad" reputation for advertising products and bugging people with those advertisements, a lot of the techniques from digital marketing apply to the work that UX writers do (Earley, 2018). Digital marketing analytics use analysis of data so they are aware of different *touchpoints* (channels) that consumers use to get to their content and how this audience can be engaged. As you learned in Chapter 5, it is important to understand your users and their touchpoints. Similarly, digital marketing analyzes touchpoints and what the user does so they can create awareness, drive traffic, generate leads, or sell something to consumers.

The same principles apply to the content and in sustaining the attention of your audience. UX writers need to keep track of the different touchpoints by which people get to your content and how to sustain and engage this audience. However, for UX writers, there is always a balance to be made between considering your users and the usability of your product and not taking advantage of them by considering the ethics of any situation. After all, we may not be directly selling users something, but we want them to come back and engage with our content.

According to digital marketing analysts, different models can capture what your users are doing. Such models are highly applicable to what you are doing as a UX writer, given that you will want to understand what your user is doing to ensure you attract, engage, and sustain them through UX writing. A number of different marketing models exist to understand the behavior of users when engaging with products that are applicable to UX writing. And because UX writing is creating and designing content as part of an interface, physical product, service, or experience, it is important that you understand these models since users will see your work as the face of the products they use. Starting from models before the internet age to today's online models, we know a lot more about the user through tracking of their online behavior (Table 8.1).

Table 8.1 Different models from marketing to understand the stages of consumer (user) behavior (Hartman, 2020).

Model	Description
Marketing Funnel (Figure 8.2)	This model is based on the linear stages of creating awareness of your brand, generating interest through marketing, and creating desire in the user that may lead to the funnel (outcome) of a purchase of a product or desired action. This model was used before websites allowed for greater tracking of the user in their journey.
Proctor and Gamble Three-Step Model (2005)	In 2005, the brand P&G created a three-step model. This model consists of a "trigger" creating a need for a user to purchase a product, and a first and second moment of evaluating the product. For example, an online ad could trigger a user to try a new product, and then run out to evaluate it in a store (FMOT), and then evaluate it a second time when they actually use the product at home (SMOT). If the first moment of truth was not a satisfactory evaluation the user would not purchase the product; whereas if the second moment of truth was not satisfactory, the user may never want to buy the product again.
McKinsey's Consumer Decision Journey (2009)	The McKinsey company identified that users go through a consumer decision journey where they *continuously* evaluate whether to use your product. After initial consideration, they go through an intense research process online before they make a purchase. They will continuously evaluate the product and reassess whether to continue with the product on their journey. Companies are wanting to capture these people in a loyalty loop so they keep on wanting to use their products.
Google's Zero Moment of Truth (2011)	Google identified that the P&G three-step model needed a "Zero Moment of Truth" (ZMOT). The ZMOT is the online behavior of users researching a product online and comparing it with others. The ZMOT yields a great amount of online data that gives insight into how users see your product as they go through multiple channels of online research (such as search engines, product pages, and online reviews). This has led to greater emphasis on digital analytics through tracking and measuring users.

Think about how you, as a person, make decisions online about products of interest. If you are like many people, you will use a search engine to find out more, read some reviews, watch a YouTube review video, and generally research the product. Now picture this same insight as needed for UX-written content. You will want to ensure that you attract, engage, and sustain your users to want to continue with your content because they will arrive at it using different routes and channels. You can no longer assume they will just find your website and locate your content, given all these channels. Instead, you have to keep track of how they come to your content and what they think of it.

How to Analyze Users' Interactions Before, During, and After Content Deployment?

Given the number of digital channels, it gets overwhelming to think about how your users get to your content and how they interact with it. Your user can get to your site's content using a search engine. They can come to your site when they click on a link on another website. They may have clicked on a link and come to your site via their social media feeds. They may have received an email or clicked on an online advertisement. So there are a number of different touchpoints that can lead them to your content. And they can also visit a number of different sites after yours. How can you make it so that your site and content are the ones where they will want to stop, spend time, and put their feet up?

As we have mentioned, UX writing focuses on creating experiences mindful of how a user interacts with them *before*, *during*, and *after*. This is necessary because people have many different options to get information and little time to waste on content that doesn't appeal to them. Here is where strategies from digital marketing analytics come in.

Various models of marketing give insight as to how to ensure your audience is engaged. A classic model of marketing is based on a funnel that identifies distinct stages from making a customer aware of your product to having them make a purchase. This approach focuses on making them aware of your brand, generating interest by advertising and fostering a desire that may lead to them performing a particular action (such as making a purchase) (see Figure 8.2).

Looking at the marketing funnel, the same applies to UX writers. That is, you need to attract them with your content, engage their interest, and sustain their desire to come back.

Now here is where it gets interesting: What if you could figure out what your user is doing and thinking *before* they take a specific action? Google's introduction of the concept of *Zero Moment of Truth (ZMOT)* specifically focuses on what a user is doing before they make a decision. Google invented this because the ZMOT is facilitated by technologies such as cookies, UTMs, and tags, and it has led to a revolution in user analytics. Site analytics and "conversion tracking" (i.e., a method to figure out whether someone performed a particular desired action) have never been more accessible to you. As a result, with data analytics, it has also never been more accessible for you to find out, sometimes in real time, what your users are doing through analysis of their online activities.

Think about this: When you were shopping for an Apple iPhone (or another brand's device), what did you do? Most likely, you used the internet to figure out whether the phone was worth purchasing. Part of this might be visiting Apple's website for new features, reading some website reviews on the phone, or perhaps watching a review video on YouTube. These types of interactions are called *touchpoints*—times when you made contact with relevant content. Visiting these touchpoints led you to perhaps compare the iPhone with other mobile phones that were cheaper and created multiple insights for you to determine

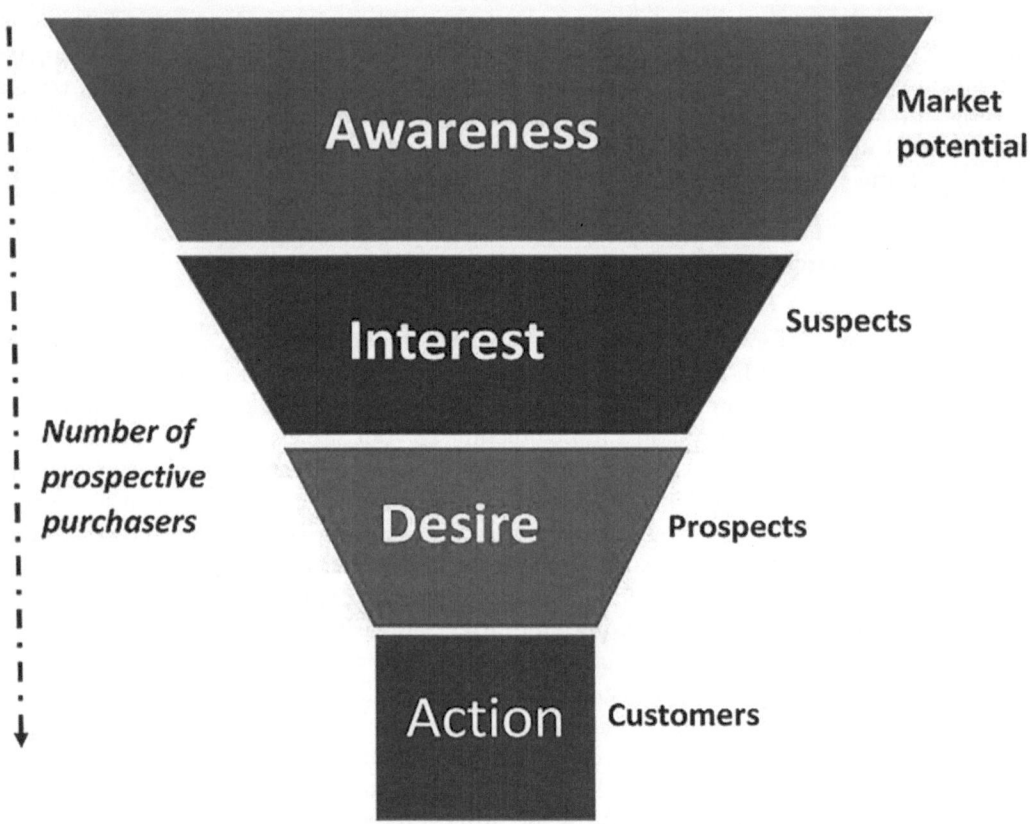

Figure 8.2 A traditional marketing funnel describes how creating awareness may funnel into desired actions by consumers.

Source: BronHiggs on Wikimedia.

whether you were interested in buying it. Now picture how companies can gain insights into what your likes and dislikes are.

According to Google's Kevin Hartman, we have more and more data about users:

> Consumers can hop between sources of information and back again as quickly as their browsers are able to load internet pages. Furthermore, consumers move between online and offline worlds fluidly—much like they live [...] Each ZMOT touchpoint represents an opportunity to interact with consumers and learn their reactions to specific brands and products. These new *touch points* provide opportunities for brands to influence consumers which requires brands to understand consumer needs—insights that could be gleaned through the collection of profile data, and online consumer behaviors.
>
> (Hartman, 2020, p. 28)

Tracking the user via touchpoint data is important because it gives insights into behavior and creates important contextual information on where users are learning more about your products. UX writers need to know how to create user personas with data and create journey (user experience) maps, making sense of which touchpoints (channels) are typically used by your users. For example, are they finding your content through a Google search only, or do they visit because you featured it prominently on the landing page or through a social media campaign? Tracking this actually sounds harder than it is.

By using a web analytics program, you can find out whether people came from Google searches on your website or the link you posted on social media. The way this works is that, when you click on a link with a UTM in it, Google Analytics (and other similar web analytics programs) are able to track that the link came from a social media campaign, a newsletter, a search engine, or your website's search bar. As a result, you can get important insights into what your users are doing before they get to your content, such as whether they happened upon it by chance or whether they were specifically looking for it on your website. Such insights give you a sense of which channels are the biggest drivers of users who will access your content.

Likewise, you can see how long users spend on your page and which links they click while there. You can also see which outside links people go to after they visit your content. Such user behavior is important to see how your content performs and can give you better ideas of your audience and what works to reach that audience. In other words, data analytics—in the form of *site analytics* and tracking of user behaviors—is important to understanding your users.

Understanding Audience and Their Behavior through Web Analytics

Through web analytics, there are ways to track what your users do *before*, *during*, and *after* they visit your content. Through web logs, you can get a pretty good idea of what users do before they get to your content and where they go after. For example, Google has your sign in with your profile through their website while you browse the web. Your sign-in allows Google to learn a couple of things about you while you visit different websites. Google can then analyze those patterns.

For example, through website traffic analysis, you can discover a number of different facts about who your user is, where they are coming from, what they do on your site, and if they are performing certain actions you intend them to do. You can get a pretty good sense of your users' demographic information and which device they use when they are accessing your content. Such tracking of the user can help you figure out how they get to your content and which channels are most successful at generating people to see and interact with your content (so-called *acquisition*). It can also tell you what people do when they are on your site and interacting with content (user *behavior*). Lastly, analytics can tell you if those users meet your daily goals of performing certain actions, such as signing up for your service or downloading a file (user *conversion*). All this information is important not only for social media influencers but also for UX writers.

Google Analytics (other analytics platforms exist as well; see Chapter 5 for content audits) can help you make sense of all these data. For example, Google Analytics can tell you how many users visited your website, how many of those were new users, and where they came from (such as a social referral from another website (e.g., clicking on a link), a direct search from a search engine, an email campaign, or a paid search where it showed up under recommend websites for the user). Digital marketers call the latter a *multi-touch attribution model*, which is important for marketers because it helps them attribute which advertisement strategies work best to drive people to content (Figure 8.3).

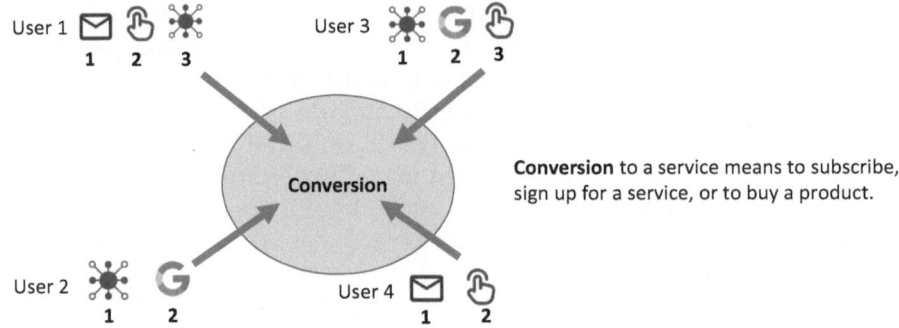

Figure 8.3 A multi-touch attribution model can help you understand how your user gets to your content.

Source: Adapted from Puri (2017).

Google Analytics is really good at providing different types of analysis of your users. However, when using such a platform, it is easy to take numbers at face value. Often, what you will find is that the numbers hide another story. Say, for example, that you are seeing a spike in visitors to a specific page. Is that spike there just because it was just freshly added and your site visitor is curious to read it, or is it because they are having a problem and looking to solve the problem by visiting this specific page? In other words, data in themselves don't say a lot, *unless you uncover the story behind it*. As you see in Table 8.2, Google Analytics can do a lot of different things, but it is up to you as a UX writer to make sense of the numbers and make them into a compelling story for other people.

Google Analytics only provides multi-touch attribution information to people who are the owners of a website. Screaming Frog, a free web crawler focusing on search engine optimization (SEO) and analysis, lets you analyze websites without having to be the owner of that site (https://www.screamingfrog.co.uk/seo-spider/).

Other platforms exist that will let you analyze different components of your content, such as:

- **Google Trends** to analyze searches for your website or brand.
- **Twitter, Facebook, Instagram,** and **LinkedIn** all feature their own analytics components that are usually proprietary and can do social media analytics.
- **Hootsuite** and **Hubspot** also feature social media analytics.

As you can tell, a lot of information can be obtained through the use of analytics. However, not every country agrees to this type of tracking, and differences in data sharing exist. The European Union has recently passed a comprehensive agreement called the *General*

Table 8.2 Using Google Analytics, different analyses can help you find out about your users and their interaction with content

Analysis	Description
Audience analysis: *Who is my user that visits my site?*	• **Demographics:** Age and gender of the audience (from their account). • **Interests:** What your audience is interested in (from their stated interests, such as music). • **Location:** The language settings and location where they access your content as specified by the browser. • **Browser, operating system:** What browser they are using and if they are on a Mac/PC. • **Device:** If they are on mobile or desktop.
Acquisition analysis: *Where is my user coming from and how did I "acquire" them?*	• **Adwords:** Paid search advertisements clicked on by your user before they got to your site. • **Search engines:** Common queries used by your audience before they landed on your page (e.g., "Book + UX Writing" or "UX writing + book"). • **Social media traffic:** Social networks used by your audience found due to them clicking on a link to you on their Twitter, Facebook, etc. • **Campaigns:** Campaigns that advertise your site and content through for instance a YouTube video link, a social media post, or email blast, all of which can draw your audience to your site.
Behavior analysis: *What is my user doing while they are on my site?*	• **Site visit frequency:** Pages visited often and for how long by your user • **Content drill-downs:** The order in which pages were visited, e.g., where your audience *landed* and where they *exited*. If a user lands on your index, checks out your company's info, and then orders something, the content drill down would show their behavior to look like this: "../index.html >../companyinfo.html >../order.html". Such content drill-downs can tell you a lot about user flows and where they typically go from one moment to the next. • **Site searches:** Searches performed by your user can tell you what they are looking for or what they may need. Search term analysis can sometimes show you if content needs to be stronger. If, for example, you see your user looking for the same information on how to install an app through your site's searches (e.g., "Instructions on how to install app"), that may mean your user cannot find it or it needs to be more prominently displayed on the page when that content is already there. • **Events:** Actions performed by your users, such as downloads, signing up for your website, logins, etc.
Conversion analysis: *Is the user "converting" (e.g., meeting a certain goal or behavior that is intended, such as signing up, making a purchase) to our organization's specific goals?*	• **Goals:** Measure specific goals, such as how many people signed up that week for your app/website and how many purchases were made per new customer vs. existing customers. These metrics can become KPIs that help your organization understand what is happening day to day. • **Ecommerce:** Measure various e-commerce transactions such as total product purchases, purchase time, and sale performance. • **Multi-channel funnels:** Measure which funnels (e.g., ad/paid search, website referral, or email or social media campaigns) drive conversion the most, say, customers signing up for your service or making a purchase. • **Attribution:** Measure which touchpoints (channels) are driving conversion so that it can be attributed to those channels. Attribution is important because it allows you to focus on, say, whether paid advertisements work well over social media campaigns.

Source: Google Analytics (2022).

Data Protection Regulation that prevents people's data from being used by companies like Google and Facebook, whereas the United States's data policy is less restrictive.

However, new privacy initiatives are taking place. Apple has rolled out an anti-tracking feature on its phones that is starting to prevent user data from being tracked. Google will not be supporting cookies starting in 2022 in its Chrome browser. Further, while most apps and websites may provide data, some apps you install on your phone are harder to track. Hence, you will find that tracking your users is not a perfect science but requires you to consider sometimes incomplete data from different channels/touchpoints subject to different rules and regulations. The area of digital marketing analytics is a constantly evolving field, so it is good to pay attention to different data-gathering practices and their impact on what you can find out about your users. In other words, finding data is only half the picture; uncovering whether the data are incomplete or limited in some way also requires you to not overstate your case. Again, uncovering the story behind your data and contextualizing it properly is as important as the data itself.

Real World Snapshot 8.1: Fabric of Digital Life Website: Where Are Visitors Coming From?

The authors of this book are involved with the Digital Life Institute (DLI). This institute is a consortium of researchers who research the effects of digital technologies on the fabric of everyday human life. As part of this initiative, *The Fabric of Digital Life* website (https://fabricofdigitallife. com) is a database of different technologies and how they interface with humans. Created by Isabel Pedersen of Ontario Tech University, the website seeks to drive interest among various researchers, including members of the DLI (Figure 8.4).

Naturally, as DLI members, we were interested in figuring out where people were coming from and how we could get more people interested in the site. To do so, and because we had permission from Isabel to access the Fabric website visitor data, we used Google Analytics to see which drove the most visits from users (Figure 8.5).

In looking at the data, we found some interesting insights into who visits the page and where they are coming from. It looks like we get a lot of Google organic searches—driving new growth with 169 new users for a total of 248 users who visit the site. There is also a lot of direct/none where it was not clear where people were coming from, in addition to direct referrals (where people click links) and searches from Bing (a search engine similar to Google). We also noticed that a "canadalearningcode.ca" page drew one user as well as a user who came from the social media platform Facebook (among other sources).

Although this gives us a better sense of where people are coming from (mostly search engines), it can also give us a sense of where we need to focus our attention if we want to drive more people to visit the Fabric website. That is, we may need to expand by having our researchers feature the link on their page to drive up referrals. Or, we can consider writing a blog post to expand the awareness of people who visit our website. Hence, tracking our users and using site analytics can help us get a better sense of where to focus our efforts and a more complete picture of who our users are and which channels they use to get to our content.

	Source / Medium ?	Acquisition			Behavior		
		Users ? ↓	New Users ?	Sessions ?	Bounce Rate ?	Pages / Session ?	Avg. Session Duration ?
		248 % of Total: 100.00% (248)	**225** % of Total: 100.00% (225)	**321** % of Total: 100.00% (321)	**66.67%** Avg for View: 66.67% (0.00%)	**2.71** Avg for View: 2.71 (0.00%)	**00:02:22** Avg for View: 00:02:22 (0.00%)
☐	1. google / organic	**176** (70.68%)	**169** (75.11%)	**205** (63.86%)	70.24%	2.15	00:01:37
☐	2. (direct) / (none)	**62** (24.90%)	**51** (22.67%)	**86** (26.79%)	65.12%	3.57	00:03:34
☐	3. t.co / referral	**3** (1.20%)	**1** (0.44%)	**14** (4.36%)	14.29%	5.07	00:05:57
☐	4. lens.google.com / referral	**2** (0.80%)	**2** (0.89%)	**2** (0.62%)	100.00%	1.00	00:00:00
☐	5. bing / organic	**1** (0.40%)	**0** (0.00%)	**3** (0.93%)	33.33%	7.00	00:10:09
☐	6. canadalearningcode.ca / referral	**1** (0.40%)	**1** (0.44%)	**1** (0.31%)	100.00%	1.00	00:00:00
☐	7. cn.bing.com / referral	**1** (0.40%)	**0** (0.00%)	**6** (1.87%)	83.33%	1.17	00:00:02
☐	8. digitallife.org / referral	**1** (0.40%)	**0** (0.00%)	**2** (0.62%)	50.00%	9.00	00:02:18
☐	9. m.facebook.com / referral	**1** (0.40%)	**1** (0.44%)	**1** (0.31%)	100.00%	1.00	00:00:00
☐	10. statics.teams.cdn.office.net / referral	**1** (0.40%)	**0** (0.00%)	**1** (0.31%)	100.00%	1.00	00:00:00

Figure 8.5 Google Analytics page of Sources/Medium for visitors to the Fabric of Digital Life website.

Figure 8.4 Fabric of Digital Life landing page.

Building Content that Attracts, Engages, and Sustains Your Users

You need data analytics to keep track of what your users are doing. You also need a way to *sustain* your audience. Consulting firm McKinsey has expanded traditional marketing insights about capturing users' behavior by introducing the "loyalty loop" (see Table 8.3). As they note, due to the availability of so many options for our audience, it is more important than ever to continuously engage your audience with rewarding experiences so they will want to come back.

Indeed, technologist Nir Eyal has written how people are "hooked" on technologies such as Google, Netflix, and Facebook nowadays because they offer a rewarding experience, leading people to want to come back to them. They do this by making the experience rewarding in some way and using an understanding of the psychology of the user's emotions and habits.

McKinsey's consumer decision journey model states that it is important not only for users to learn about your business online but also to continuously want to be rewarded as part of the "loyalty loop." Users, once they decide on your products or services, continuously evaluate whether they see them as valuable.

Loyalty loops play a large role in getting people to want to continue with your content. Think about how streaming service Netflix's smart design of auto playing the next episode of your favorite television series takes the task of clicking on the next episode out of your hands. Instead, you get a notification and a countdown of when the next episode automatically starts to play (usually in 10 seconds). This feature is so popular that it has entered the public consciousness with the term "binge watching." Or think about how Amazon.com stores your payment information after your first order, so whenever you order again, you needn't spend a lot of time entering your payment information or your address again. These are UX writing decisions that make the user's experience more favorable, but they work so well that they have the innate effect of making people part of a subscription model. However, as you've learned throughout this book, as a UX writer, it is of course your responsibility to be ethical in designing and balancing the product's needs with the user's behavior.

Table 8.3 Consulting firm McKinsey's consumer decision journey describes how users seek rewarding experiences as part of a "loyalty loop," making it more important than ever to continuously engage them with high-quality content

Phase 1: Initial Research & Consideration	*Phase 2: Moment of Purchase*	*Phase 3: Loyalty Loop*
The user researches and actively seeks information on a product (shopping websites, reading reviews, watching YouTube videos).	The user decides whether the product is worth it.	The user becomes a customer, receives rewards from using the product, and decides to stay with it due to ongoing positive triggers post-purchase (such as new content and perks offered as a result of being a subscriber).

Source: Adapted from Court et al. (2009).

Real World Snapshot 8.2: Thrown for a Loop: Using the Consumer Decision Journey Model to Create Rewarding Experiences for Users of a Second-hand Clothes Selling App

Madelyn Simmons works for a company that has created an app, Loved Vintage Clothing (LVC, pronounced "Lovesy"), that helps people sell their gently used clothes and search for popular brands. In analyzing their traffic, she noticed that many people tried out the Lovesy app only once but failed to continue to use it. She also knew from analyzing web traffic that a lot of people were coming in from campaigns they ran on social media. She also noticed that despite their CEO's insistence on the heavy promotion of the app on used clothing web forums, this wasn't driving people to "convert" and download their app. In other words, the company knew where its users were coming from, but when they did, people weren't coming back. It was throwing everyone at the company for a loop. What was needed was a "loyalty loop" that got people to notice the app, get excited, and want to return.

Madelyn had ideas about making things more exciting for users so they would return to the LVC app. She knew that poor advertising was not leading people to try out the app. She decided to use the *four stages of the consumer decision journey* to map out what *ought to happen* when a potential user notices the LVC app and to present this to their CEO.

First, with the help of her teammates, Madelyn started mapping out who the other players were in the used clothing demographic. She did this so that she could tell what users were taking for their initial consideration set. In other words, she wanted to know the initial set of brands that popped up in their heads or what they found online when looking up "second-hand clothes selling app."

Second, Madelyn started looking at how users *evaluated* their app on social media, in online comments, and on message boards. She noticed a lot of people were saying the LVC app looked cool but didn't excite them. She also noticed many comments about other apps offering "power-seller" ratings on sellers in the community. Users appreciated these features in the other apps because they inspired more confidence in sellers in the community.

Third, she looked at and evaluated the other apps and how they got users to "convert" to using their services. Madelyn noticed that for many users, signing up relied on seeing the app as a social network that sold clothes. Simply put, the social network feature helped them see who was active in the community and maintained their profile, which inspired trust in buyers and other sellers. Many companies have developed this aspect of their app's UX for their users, which LVC was missing.

Last, Madelyn knew the app needed a "hook"—something users really liked. She knew that other apps had social network features, but what was missing was a credit system with discounts and effective mobile integration. To make people want to return and keep using LVC, Madelyn thought it would be a great idea to send out encouraging mobile in-app messages directly to the user with discounts ("we noticed that you purchased; here's a 10% discount on your next purchase") together with special promotions ("deals on jeans 20% off only this week!).

Madelyn's experience can help us see how understanding the decisions users go through in their decision journey can inform our practices as UX writers. It helps UX writers build better user journey maps by not only infusing the maps with factual data obtained from data analytics but also by incorporating information about triggers and hooks. Using these data enables UX writers to build repeated user visits, desirable user experiences, and brand loyalty.

Using Key-Performance Indicators (KPIs) to Measure Content Performance

So far in this chapter, you've learned about how a content creation framework may also require you to track how your content is performing with the user. The use of digital analytics, particularly site traffic analytics, can give you a better sense of who your users are, which channels they come from when they visit your content, how they navigate your site, and if they perform tasks that you want them to do (such as downloading a file and signing up for your service). And you have also learned about different models from digital marketing about how to attract and engage your audience with loyalty loops. One of the missing pieces here is how to measure, on *an ongoing basis*, what your user is doing and make a case for the value of your work as a UX writer.

For that, you will need to develop *metrics that help you keep track of content performance*. Simply put, metrics are measurements that help you understand what is going on with your users and content. Metrics are important in modern organizations because they help you make arguments about how the user or your content performs in relation to the organizational goals. A metric is something that you measure.

An important distinction must be made here between a *metric* and *a KPI*.

- **Metric:** as the name indicates, it measures a specific aspect of the user and creates a score. A metric can measure a score your user gives for your content, how long they stay on your page, or how often they access your content.
- **A KPI:** a KPI is a *goal* specified by your organization that can also be measured and expressed through a metric. A KPI is broad and can involve increasing revenue, efficiency or user performance, or any other goal defined by your organization. For that reason, KPIs are often shared with different stakeholders, such as managers and different people in a team, so that they can measure how this goal is being met or not met.

The difference between KPIs and metrics is that a metric captures a specific aspect of the user's experience that is important to you, whereas a KPI is shared in an organization as an important metric for the well-being of the company that may involve the user or other goals. As their name indicates, KPIs *indicate* how the organization is *performing* in *key* areas vital to the organization, and as such, they are more broad and spread across the organization. In other words, KPIs are really important for making the value of your work known.

Think of KPIs as the indicators on your car's dashboard. When you are driving, you are always checking your vehicle's speed and whether you have enough gasoline. You use these metrics (because that is what they are—measurements) to make decisions from one moment to the next. Similarly, companies use KPI dashboards with performance data to indicate how they are performing and to make better decisions.

Some commonly used KPIs are:

- **Revenue:** How can we make more profit?
- **Cost:** How can we reduce costs?
- **Performance:** How can we increase performance?
- **Efficiency:** How can we increase efficiency?
- **User satisfaction:** How can we increase user satisfaction with us?
- **Usability:** How can we increase the usability of our site for our users?
- **Other:** Any other goal that you have that can be measured.

As a UX writer, it is therefore important that you know which metrics describe what the user is doing and how your metrics relate to the KPIs of user performance, efficiency, or user satisfaction (next to revenue, cost, and any other goals). If you are writing content for websites or digital products, you will probably want to focus on the KPI of the System Usability Score (SUS), which we covered in Chapter 7 (see Figure 7.3). Remember, unlike academic grades, a good SUS score is a 70.

Why is this important, you ask? Organizations may not always know what you contribute (they may think you are "just a writer"), but if you make a case of why what you do connects to their KPIs, then you are also giving them a good sense of return on investment (ROI). Unfortunately (or fortunately), organizations are always concerned about the *bottom line*. What you do as a UX writer, as the first point of contact between the user and the business, can help them understand the value of what you bring as a UX writer by connecting metrics to KPIs and explaining how your work has a good ROI.

How Can You Use Metrics and KPIs in UX Writing?

Metrics are measurements made through data. And you know that these metrics can also be part of a KPI that an organization adopts to assess its performance. The old business saying "you can't manage what you can't measure" demonstrates the importance of data in not only the modern workplace but also in the life of a UX writer.

However, that doesn't mean the data you use to measure your metric are always correct. If your data are incomplete, don't apply to your user or content, or simply are incorrectly analyzed, they are virtually worthless as a metric or KPI. Hence, data are never the only answer to get a clear idea of what your user is doing. You will still need to contextualize what is going on with your user or consider errors of interpretation. For example, if someone stays only a short time on a page, is that because they made an immediate decision that the page wasn't helpful or because they were able to find answers quickly? Similarly, if someone stays on a page for a long time, is it because they were having a hard time finding what they need or because they found the content on the page useful and wanted to read it carefully? In other words, time-on-page needs to be complemented with information about your users' actual experience—which you can get from probing them for answers to "why?"—in order to make accurate observations about content performance.

Data, through metrics, are only part of the picture, and sometimes you will find that other data need to be gathered or other metrics need to be adopted by your organization. Or you will need to interview your users to find out aspects that simply cannot be captured through a quantitative metric. This is important because we know UX design involves people's emotions and feelings, which may not always be fully captured through cold numbers and may also require you to interview them to collect qualitative data (Verhulsdonck & Shalamova, 2020). In other words, quantitative metrics aren't the only way to find out about your users, but they are often important in a business setting.

Beyond business goals unique to your organization, Heap (2022) identified five specific types of metrics that help you think about and understand the user and your content. The five metrics are: (1) *breadth*, (2) *depth*, (3) *usability*, (4) *frequency,* and (5) *sentiment*. Using these five types of metrics lets you think about different types of user interactions with your content and is therefore useful for benchmarking (see Table 8.4).

Like it or not, UX writers have goals to meet. As a UX writer, you will, therefore, most likely also have to consider KPIs to meet the business objectives of your company or organization. Perhaps your company asks you to grow the number of users who access your content, make a site or app more navigable, or simply increase the number of users who perform a specific objective (such as signing up for your service or downloading your app or a file). The combination of how your content is structured, whether the text is accessible and understood, and whether your user is attracted and engaged all play an important role in UX, but these can also be measured by KPIs to make decisions on what changes are crucial (Table 8.5).

Table 8.4 Using different metrics gets you to understand different dimensions of your user space and their interactions

Metric	*Typical Metric name / how to calculate*
Breadth – Measure how many subscribed users vs. overall users make use of your content/product.	• **Adoption rate:** Specific users divided by total number of users. • **Scenario:** One can calculate, for example, how many users adopted using a specific feature in your content by dividing this number by the total number of users. An adoption rate metric can tell you if you are meeting your goal on a weekly basis and how much breadth among your users the feature has (e.g., how many of the total users have embraced the feature).
Depth – Measures the number of specific features an average user accesses/uses.	• **Rate of use:** Percentage of users from total users who use say, three or more pages in your content, or how many daily, weekly, or monthly active users use a particular feature. • **Scenario:** One can use the rate of this metric to look at how much your users access content or how frequently they use it. This gives a UX writer insight into the *depth* of the user's behavior when interacting with specific features to make a determination whether it works or whether particular features need to be changed. It also helps to segment your users and what they do (e.g., you get insight into what your daily, weekly, and monthly users do and the level with which they engage with your content).
Usability – Measures how much effort is needed by your user to accomplish a specific task.	• **Time-on-Task:** Average time users spend on completing a task. • **Scenario:** To see if new instructions work better for accomplishing a task, you can measure the time-on-task of your user. If you have established a particular time-on-task as a baseline for doing the task and you have released new instructions and the average time-on-task decreases, that means the instructions work well. Do take note if you are using time-on-task as an efficiency metric; a decrease signals you are doing well. However, if you are using it as an engagement metric, decreased time-on-task is not necessarily a good sign because your users are less engaged.

(*Continued*)

Table 8.4 (Continued)

Metric	Typical Metric name / how to calculate
Frequency – Measures how often your user does something.	• **Number of Daily/Weekly/Monthly Visits:** You can measure how many people visit your site and spend significant time on it. • **Scenario:** To see if, after deploying new content, your user likes this content, you can measure whether this increases the number of visits. Take note that if there is new content, people may read it, and your number goes up. What you want to measure is over time; whether this sustains.
Sentiment – Measures how your users feel about your content/ product.	• **Net Promoter Scores (NPS):** By having the percentage of people that like your product or service and would promote it, and subtracting the percentage of people that would actively discourage people from using it, you get a score from –100 to 100. The NPS score describes those that actively dislike (–100-0 score), are neutral (1–30), think it is decent (31–50), good (51–70) to excellent (71–100). • **Scenario:** If you want to figure out if people liked their experience with a product, you can survey them after using a Single-Ease-Question, "On a scale of 0–10, how much would you recommend our product to your friends?" which yields how they feel.

Source: Heap (2022).

Table 8.5 KPIs by content type differ based on the medium and goals of the organization for deploying content.

Blog posts/ articles	Email	Social media	Videos	Podcasts	PPC campaigns
Web traffic	Open rate	Amplification	Views	Subscribers	Cost per click
Unique visitors	Conversion	rate	Unique	Backlinks	Click-through
New vs.	rate	Applause rate	viewers	Downloads	rate
returning	Opt-out rate	Followers/fans	Avg. view	Social shares	Ad position
visitors	Subscribers	Conversion	duration	Reviews/	Conversion
Time on site	Churn rate	rate	Subscribers	ratings	Conversion
Avg. time on	Click-through	Landing page	Impression		rate
page	rate	conversion	click-		Cost per
Bounce rate	Delivery rate	rate	through rate		conversion
Exit rate		Return on	YouTube		Cost per sale
Page views		engagement	CTA click-		Return on ad
Page views per		Post reach	throughs		spend
visit			Shares		Wasted spend
Traffic sources			Comments		Impressions
Geographic			Traffic sources		Quality score
trends					Total spend
Mobile visitors					
Desktop visitors					
Visits per					
channel					

Source: Adapted from the Content Marketing Institute.

The use of KPIs is prevalent in many companies and organizations because they can get near-real-time data on what their users are doing when they are interacting with content. KPIs are often created by the organization and disseminated as a means to discuss the objectives of that organization in an ongoing manner. This requires understanding what you aim to measure and how this relates to the overall goals of your organization.

Adopting KPIs to Frame Your Content for Actionable Insights

Often, the goals of an organization can be quite broad. They are looking to increase their user base, or the number of people that view their content, in hopes that they will come back. They may look to expand their advertisement revenue and attract more eyeballs. Usually, if it is a company, they are looking for ROI: more subscribers, higher advertisement revenues, and more shares. However, here is where it gets tricky: does having more site visitors automatically translate to a better experience for the user? Not necessarily. You will need to use KPIs and reflect on how they relate to your organization's goals and the overall user experience and quality of your content.

The ability to attract more visitors to your site doesn't always translate into a good ROI if you don't understand whether something works as a KPI. Many organizations have the goal to get more first-time visitors. As the logic goes, the more people who interact with your content, the better, right? Not necessarily. You will need to determine whether or not adopting a specific intervention makes sense and think about how it may impact your KPI.

For example, you may have adopted a KPI goal of increasing "new site visits," where you use a benchmark to measure how many more new users visit your content. Let's assume your organization is on board with this. Once you deploy your KPI, you notice many new users visiting your site. This is great news, right? But while you are looking into what these new users are doing, you notice the *bounce rate* is horrible. The bounce rate is a metric that expresses how quickly your user bounces—meaning leaves—after they are on your page. You find out your new users are literally spending a few seconds on the website before they leave. You are worried that your measures reveal your site may be good at *attracting* new users, but you are not good at *sustaining* them with content—they are literally leaving after a few seconds without engaging with your content.

You could employ the KPI of "new site visitors" and present it to your colleagues as a great success. More than likely, however, you recognize your that "new site visitor" KPI is a *vanity metric*. A vanity metric sounds great, but it doesn't really tell you a whole lot about what needs to be done to improve things. Companies and organizations usually go for vanity metrics because they are easy to understand—total visits per month, new visitors. However, having high visitor counts may mean your marketing works, but these same users leaving really quickly is not good news since your work as a UX writer means that you are tasked with writing engaging content for your users.

Instead, you need an *actionable metric*—a metric that gets your organization to understand that this is the information they need to act on right away (Verhulsdonck & Shah, 2021). Instead of just presenting your high number of new site visitors, you decide to present the high bounce rate to your colleagues as an important KPI metric. This metric is actionable and can lead to you adopting a new goal: reducing the bounce rate by improving the user experience. In turn, you can then introduce other KPIs to your organization that help capture and understand what your new site visitors want to see. Using measures such as on-page user surveys and having the goal of improving the overall user score on whether content met their expectations will give you a better sense of where content is performing well and where it is not. Using KPIs requires you to think about your organization's goals in an ongoing manner and to identify whether they work for the objectives of your organization (and if they are actionable!).

A UX Metric Framework: The 3×3 Method

In this section, we offer a method of thinking about your users using a framework and metrics. We call this the *3 by 3 User-Metric Method*. There are, incidentally, many 3×3 methods in UX. Our 3×3 User-Metric Method consists of:

- **3× Identify:** indicate specific ways to (1) *state your goals*, (2) *identify which channels* let you measure these goals, and (3) *specify which metrics to use* to capture what users are doing.
- **3× Measure:** consider metrics as expressing (1) *attitudes*, (2) *behaviors*, or (3) *tasks* to measure what is going on.

Table 8.6 presents this method as a matrix.

This 3×3 Method borrows from Google's HEART framework. Google has developed a framework to think about the user in relation to content using common KPIs. The HEART framework is great in letting UX writers think about the threefold process of connecting their work to specific goals, knowing where to gather data, and expressing whether goals are met through specific UX metrics. The framework is therefore useful to UX writers as well.

If you are trying to specify your goals, think of this in terms of Google's HEART acronym, which focuses on the following UX goals:

- **Happiness:** How happy the user is with the content and whether they recommend it.
- **Engagement:** How often users engage with content.
- **Adoption:** How many new users adopt your content.
- **Retention:** How many users re-subscribe and keep coming back to your content.
- **Task Success:** How easily and quickly a user completes a key task (Table 8.7).

To use the HEART framework in your work as a UX writer, think about writing out the following three components in order: Given that you have to start with a goal, you can then work your way backward into where to capture data and express these in specific metrics that people understand.

- **Stage 1: Goals:** *What do you intend to happen to your user, and how can you express that in terms of a specific broader goal or KPI?*

  - Think of goals as the increase/decrease of certain characteristics of the user and their experience. A goal, then, could be an increase in users who access your content, sign

Table 8.6   3 by 3 User-metric method

1 Goal	2 Signal	3 Metric	Which of these 3 metrics apply
What do you intend to happen to your user, and how can you express that in terms of a specific broader goal (KPI)?	*Which channels (web, mobile, social media, and apps) allow you to measure what the user does that represents that goal?*	*How can you measure and reflect the user and their ongoing experiences through a specific metric?*	*Can your metric be captured in one of these three major metric categories?* 1 *User **Attitude** (how the user feels about something)* 2 *User **Behavior** (how the user acts)* 3 ***Task** (in what order the user performs a task)*

Table 8.7 The HEART framework helps researchers understand what is being measured through different metrics.

	Goals	*Signals*	*Metrics*
Happiness	Users find the app helpful, fun, and easy to use	• Responding to surveys • Leaving 5-star ratings • Leaving user feedback	• NPS • Customer satisfaction rating • Number of 5-star reviews
Engagement	Users enjoy app content and keep engaging with it	• Spending more time in the app	• Average session length • Average session frequency • Number of conversions (consuming content, uploading files, purchases, etc.)
Adoption	New users see the value in the product or new feature	• Downloading, launching app • Signing up for an account • Using a new feature	• Download rate • Registration rate • Feature adoption rate
Retention	Users keep coming back to the app to complete a key action	• Staying active in the app • Renewing a subscription • Making repeat purchases	• Churn rate • Subscription renewal rate
Task success	Users complete their goal quickly and easily	• Finding and viewing content quickly • Completing tasks efficiently	• Search exit rate • Crash rate

Source: Adapted from Bonnie (2021, n.p.).

up for your service, or say they are satisfied with your content. Or it could be a decrease in the time they spent reading instructions and doing a task more efficiently.

- **Stage 2: Signals:** *Which channels (web, mobile, social media, and apps) allow you to measure what the user who represents that goal?*

  - Think about the different touchpoints your user interacts with and leaves traces of interactions. Those channels also let you track signals from the users interacting with your content, such as how many sign up daily, who filled out a satisfaction survey, or how long they spent on a certain page. Signals can come from, say, a social media campaign or an email you sent out that drive changes in what users are doing.

- **Stage 3: Metrics:** *How can you measure and reflect the user and their ongoing experiences through a metric?*

  - Metrics are the analysis of signals from different channels. These come from analyzing the behavior of users, e.g., what they are typically doing in an app or website you developed, which content they access, for how long they stay, etc.

To help yourself create metrics as part of the 3×3 Method, think about your user in terms of expressing *attitudes*, *behaviors*, and *tasks*.

- **Attitudinal metric:** measures the user's attitude (feelings and emotions) toward your content. You are measuring the feelings of the user in relation to your content, e.g., their level of satisfaction with the usability (SUS) or whether they would say positive things about you (Net Promoter Score (NPS)).

- **Behavior metric:** measures the user and their behavior in relation to your content. You are measuring how your user is behaving, e.g., are they downloading a file you offer on your page (a so-called conversion), are they stopping subscribing to your page (e.g., a churn rate), or are they continuing to do so (e.g., renewals).
- **Task/Frequency metric:** measures if a task was completed, how long this took, and the frequency of when something occurred. You are measuring how many times a task was attempted and how long this took, e.g., task success rate, task completion time, etc.

Validating Your Successful Metrics

You have learned that a metric is just a measurement and needs to be meaningful and actionable for your organization. You have also learned that for metrics to work and for your organization to see the value of your work, you have to connect them to a KPI like performance, efficiency, satisfaction, revenue, or cost. You have also learned how to use the 3×3 Method to think about this concretely by identifying goals, signals, and metrics and expressing this in attitudinal, behavioral, or task metrics.

However, one thing that is missing is that you need to understand how your metrics stack up against your goals. You will want to make it clear how your data stacks up against established standards for your work. For that, you need benchmarking. Benchmarking is the process of comparing yourself to other competitors. This lets you *validate* your findings and compare your data to industry standards and common user expectations.

What Is Benchmarking and How to Validate Measurements?

Many businesses use *benchmarking* to compare how they are doing with a standard. A benchmark is an overall goal to compare current performance. Often, a benchmark is set based on either past performance metrics or the benchmark standards for a particular industry. For example, perhaps you've started a metric that measures a specific time you think a user can make a purchase in your e-commerce system. Using this metric as a benchmark, you can make a change in your e-commerce system interface and then measure whether people take *more* or *less* time to make a purchase compared to the older interface to see if you are improving.

Alternatively, there are also external benchmarks set per industry. For example, the NPS measures how many people are actively promoting your product or service versus those that do not like it. Perhaps you've already been asked this when you visited a website or used a particular online service: *"On a scale of 0–10, how likely are you to recommend us to a friend or colleague?"* What is behind this question is the product owner's desire to calculate the NPS to ascertain whether or not their users will recommend them to others.

Generally, an NPS score is between –100 and 100. Organizations want to know whether they have loyal users (promoters who have nothing but positive things to say) and whether those *outweigh* the people that say negative things about them (detractors or people who are not enthusiastic or downright negative about their product or service) (Figure 8.6).

Starting with the percentage of people who are promoters (people who have positive feelings toward you) and subtracting the percentage who are detractors (people who scored you as six or less), you will then get a number from 100 to –100. There is also a group of passives (people who score you seven to eight), but passive scores are left out of this metric.

Monitoring the NPS metric lets you measure whether people are still loyal to your brand's experiences with a concrete benchmark that is easy to understand for everyone in your

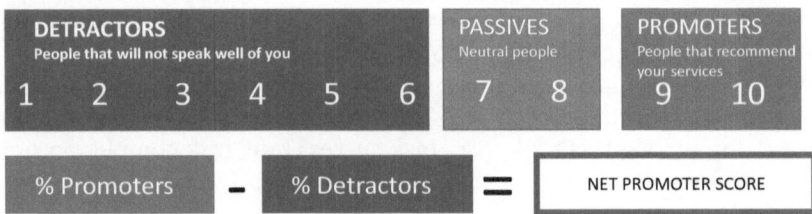

Net Promotor Score:

How likely is it that you would recommend Company X to a friend or colleague?

Figure 8.6 The Net Promoter Score calculates the percentage of people who rate your site with a 9 or 10 score (Promoters) minus those that rate it 6 or below (Detractors).

Source: Adapted from Qualtrics (2022).

Table 8.8 A Net Promoter Score is a benchmark metric that lets you figure out if more people promote your product or service than speak negatively of it

Net promoter score	How your company is doing
–100-0	• A majority of people do not like your product or service. Your customer UX really needs improvement or you risk losing these users.
1–30	• More people than not like your product or service. There is a lot of room for improvement. Your customer UX can be improved.
30–50	• People are generally happy with your services and are okay with promoting them. Your customer UX is generally good.
50–70	• Larger than average people are very happy with their experiences with your services. Your organization is doing really well in customer UX, and your users are happy with you.
71–100	• This is the gold standard and it's very difficult to obtain. Your company is the best at customer UX, and you have users who are loyal to you.

Source: Delight (2022).

organization. Take note that certain industries have different NPS benchmarks (average scores). For example, some companies are more negatively perceived than others and thus may have a lower NPS on average as a benchmark. For example, an airline company often has a lower NPS because booking an airline ticket, getting on a plane, and flight delays or cancellations create more frustrated customers. In comparison, the NPS for, say, a coffee company like Starbucks may be higher due to the consistency and simplicity of the process of getting a beverage (Table 8.8).

To validate what you are doing, you can use benchmarks and comparisons with earlier benchmarks (metrics you deployed earlier, NPS, or other metrics).

Sustaining User Engagement through Ongoing Measurements

Next to validating your findings with established benchmarks, you will want to sustain your user base by measuring on an ongoing basis what is happening. Think about ongoing measurements as consisting of you checking in on your KPIs and various metrics that you employ through, say, Google Analytics and other analytics programs that capture what your user is doing. You can have your dashboard capture how many people visit your site. A crucial component to differentiate here is between *real-time analytics* and near-real-time versus downloaded results. Depending on your work setup, you may have real-time analytics if your company has set up a good analytics platform that lets you find out what your users are doing at that exact moment. Sometimes that is not always possible, and you may get near-real-time (perhaps a daily tally) data. Other times, you may not be able to get real-time data and may have to ask your system/site administrator, or analytics team to download data for you that they share at an appropriate time.

Needless to say, you will want to make sure that you engage and sustain your users' interactions with your content by continuously measuring how your KPIs are performing. For example, you may notice that a popular article on your site is underperforming relative to when you first deployed it. This could lead you down the path of measuring what articles are performing well. You could then do a SEO (search engine optimization) analysis to see what keywords users use in their search engines to get to your site as a means to figure out the evolving needs of your users. Similarly, you could use Google Analytics to see how people progress through the site through common user webpage flows and make inferences about what is popular. In turn, such ongoing measurements help you think about new ways to meet your users' needs and develop new content and approaches.

Conclusion

In this chapter, you have learned about how common content creation frameworks require you to keep track of your users in an ongoing manner. You have learned how to *attract, engage,* and *sustain* your users through content that funnels their interest and sustains it through various stages of engagement in their consumer (user) decision journey using loyalty loops (engaging content and value). As part of this, you learned about the ZMOT, now used by companies like Google to track what users are doing before, during, and after an interaction. You have also learned the value of UX metrics—specific measurements—and KPIs that are broadly shared goals in your organization to help you think about your users and content while promoting the value of your work to that organization.

In this chapter, we also introduced the 3×3 Method of identifying your goals, signals (channels), and metrics and seeing those metrics in terms of your users' attitudes, behaviors, or a more general task perspective. This method will give you a framework for establishing how to track and measure what is important. Last, we introduced the importance of benchmarking against established industry standards and doing so in an ongoing manner to provide relevant content and validate insights so you can sustain your users.

Chapter Checklist

This chapter has covered how you can track and measure the success of your content, which is important in today's attention economy where users have many competing options. You want to make sure you attract, engage, and sustain your users by closely tracking and measuring how your content performs.

- *Create a content creation framework to plan out how you are going to attract and engage your users.*

  - Develop a plan for what content needs to be developed and a timeline for when this content is distributed to your audience, and develop a workflow as to who will create what content (if working in a team).
  - Use a spreadsheet to keep track of who will review and edit the content so it follows guidelines for voice and tone, and measure how many people liked, commented on, or

shared your content on a frequent basis (daily, weekly, or monthly) to keep track of how your content performs.

- *Use tracking technologies to see how your content performs and what users do when they encounter it.*

  - Develop a particular marketing model to frame how to attract, engage, and sustain your user, such as McKinsey's consumer decision journey or Google's ZMOT, to model how your user engages with your content in an ongoing manner.
  - Make use of a web analytics platform such as Google Analytics to keep track of your users using this model and to develop metrics.

- *Distinguish different sources that allow your audience to reach your content.*

  - Use and distinguish between social referrals from other sites, direct vs. paid search engine results, and email and social media campaigns.
  - Develop a multitouch attribution model to see that yields the most visitors, and develop your strategy based on this segment of your audience to strengthen how you sustain this audience.

- *When using any analytics platform (Google Analytics or others), make sure to distinguish different units of analysis that let you determine what is happening.*

  - Distinguish different analytics based on *audience* (who is visiting my site), *acquisition* (where is my audience coming from), *behavior* (what do they do on my site), and *conversion* (are users converting to specific goals that I want them to perform, such as signing up for my website).
  - Remember that different regions (e.g., Europe, Asia) and companies (Apple) have different privacy regulations, which may impact the types of data you can expect to get from these entities.

- *Create metrics that mean something to your organization.*

  - Distinguish between KPIs, which are stated metrics key to your organization, and regular metrics.
  - Embrace actionable metrics that give an idea of what actions your organization can take based on the story they tell; avoid vanity metrics that look good but do not tell the story of your audience.

- *Use the 3×3 UX User-Metric Method to quickly develop a better understanding of your user and which metric you need to adopt.*

  - First, following Google's HEART model for metrics, identify (1) the goals of your organization, (2) which signals from specific channels can capture those goals, and (3) embrace specific metrics to capture what users are doing.
  - Second, choose what you want to measure; e.g., your metrics should express the (1) attitudes, (2) behaviors, or (3) specific task(s) that you want the user to complete.

- *Validate your measurements through benchmarking so your metrics are meaningful given the preceding performance.*

  - Benchmarking lets you apply industry standards to what you can reasonably expect as a standard task duration to see how you measure up.
  - Use the NPS to measure how people rate your company overall and compare it to others in the same industry segment.

Discussion Questions

1 Why is a content creation framework necessary?
2 What is the concept of Google's ZMOT, and what impact has this had on analytics that could be used for UX writing purposes?
3 What distinguishes a KPI from a metric? Why is a KPI important in a business context?
4 Explain the difference between *acquisition* and *conversion* in an analytics context. Why are these important to measure in terms of the user?
5 Explain, in your own words, what does the 3×3 User-Metric Method do?

Learning Activity

Find a popular website you use a lot. Using your own knowledge of the site, be sure to distinguish at least three channels you think your user may be coming from to the site. For instance, perhaps they send out updates via social media and draw people like yourself to their content. Then, establish a goal for your user and develop a method of measuring it based on the "KPIs by content" visual in this chapter. This means you will need to research these KPIs and what they mean. Then, remember that validation requires you to make repeated measures and benchmark them, make a case for how you will measure, and create a benchmark so you can validate your findings.

References

B2B International. (n.d.). What is employee NPS? https://www.b2binternational.com/research/methods/faq/what-is-employee-nps/

Beck, E. N. (2015). The invisible digital identity: Assemblages in digital networks. *Computers and Composition, 35*, 125–140.

Bonnie, E. (2021). "How to use the Google HEART framework to measure and improve your app's UX". *Clevertap*. https://clevertap.com/blog/google-heart-framework/

Court, D., Elzinga, D., Mulder, S., & Vetvik, O. J. (2009). The consumer decision journey. *McKinsey Quarterly, 3*(3), 96–107.

Delighted. Experience management 101: What is a good NPS score? https://delighted.com/blog/what-is-a-good-nps-scoree

Earley, S. (2018). AI, chatbots and content, oh my! (Or technical writers are doomed—to lifelong employment). *Intercom, (65)*1, 12–14.

Eyal, N. (2014). *Hooked: How to build habit-forming products.* Penguin.

Google Analytics. (2022). *Google analytics for beginners.* https://analytics.google.com/analytics/academy/course/6

Hartman, K. (2020). *Digital marketing analytics: In theory and in practice.* Kevin Hartman.

Heap. (2022). 5 Core Metrics: A framework for measuring product success. [White paper]. https://heap.io/resources/ebooks-whitepapers/5-core-metrics-framework

Hubspot. (2022). *Content marketing certification course.* Hubspot.com.

Puri, S. (2017). Understanding multi-channel attribution modelling – Demystifying your attribution data. https://www.adalyz.com/multi-channel-attribution/

Qualtrics XM. (2022). Net Promoter Score (NPS) Question. https://www.qualtrics.com/support/survey-platform/survey-module/editing-questions/question-types-guide/specialty-questions/net-promoter-score/

Verhulsdonck, G., & Shah, V. (2021). Lean data visualization: Considering actionable metrics for technical communication. *Journal of Business and Technical Communication, 35*(1), 57–64.

Verhulsdonck, G., & Shalamova, N. (2020). Creating content that influences people: Considering user experience and behavioral design in technical communication. *Journal of Technical Writing and Communication, 50*(4), 376–400.

Part 3

Practices

In this last section of the book, we explain the common forms of UX writing and give you guidance on creating your own portfolio. We also include the pressing concerns about using automated content generators such as generative artificial intelligence (AI) to create UX content. **Chapter 9** goes over six popular UX writing genres and their associated tasks and challenges. You will examine the forms, structures, and delivery of these content types. In **Chapter 10**, you can expect to learn tips and tools for building an attractive UX writing portfolio with confidence. We will show you how to apply the lessons from this book to land a career in UX writing. Because AI was such a prevalent issue during the writing of this book, we have given it special attention and a detailed investigation. **Chapter 11** provides a thorough overview of the functions and recommended uses of generative AI. **Chapter 12** closes with sample scenarios for using AI in UX writing tasks.

DOI: 10.4324/9781003274414-11

9 Popular UX Writing Genres and Tasks

Chapter Overview

This chapter specifies the expectations for user-centered content in six popular UX writing genres. Here, we showcase common tasks involved in these UX writing genres and give readers a close look at the forms, structures, and delivery of content. You will learn the dos and don'ts of designing these content types and practice applying these tricks through some of the embedded exercises in the chapter.

Learning Objectives

- Differentiate between microcopy and microcontent.
- Articulate the primary goals and features of the six popular UX writing genres.
- Develop writing strategies to address common challenges in UX writing tasks.
- Design user-centered content via texts as well as videos and voice interfaces.

UX Writing Products

Throughout this book, we have included a plethora of UX writing examples and situations. You may recall the flight delay messaging challenge in the opening of Chapter 3 or the website content design project in Chapters 4 through 6. Perhaps you will find the metadata and browser cookie analytics in Chapters 7 and 8 most intriguing. All of these instances are considered products of UX writing. As we have emphasized from the start of the book, writing for user experience is to produce content and design that care about user needs and requirements, whether through front-end elements (like website copy and product catalog), back-end components (site analytics, application programming interfaces (APIs), and software documentation for developers), or writing that appears *around* the interface (such as icons, nudges, system feedback, and chatbot conversations).

Given the complexity of digital writing and the contextual nature of UX writing today, you may find yourself responsible for designing any of the above products for your team and company. Depending on the project you're on, you could be the researcher, the content developer, the product tester, or the project manager who works as a content liaison with engineers. No matter the role, your understanding of the general genres and tasks in UX

DOI: 10.4324/9781003274414-12

writing would greatly benefit you in the world of content design. In this chapter, you will learn the six most prevalent types of UX content that writers engage with:

- Onboarding experiences
- Help guides and contextual tool tips
- Error or system messages
- Forms and labels
- Legal notices
- Settings and specs

Microcopy and Microcontent

Before we jump into the common genres of UX writing, let's unpack one of the most used words in UX writing that serves as a catch-all term for user-interface content: Microcopy. We have refrained from referring to microcopy as a genre on its own because it's too broad in coverage. As we demonstrate in the following sections, every type and task of UX writing involves microcopy, whether it is textual, graphical, audio, tactile, or a combination of these modalities. Figure 9.1 shows a compilation of these microcopy examples. Take a look and discuss these with a friend: Which of the microcopy examples do you come across most in your daily interactions with your technologies? Which are less common?

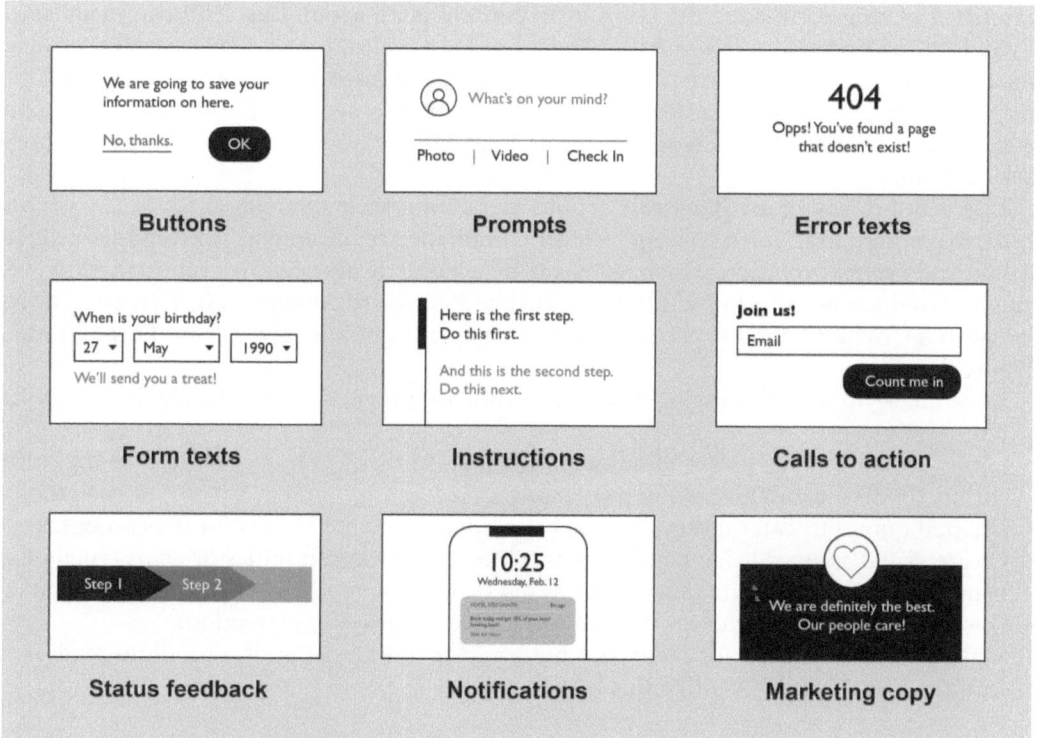

Figure 9.1 Popular microcopy examples found in various user interfaces.

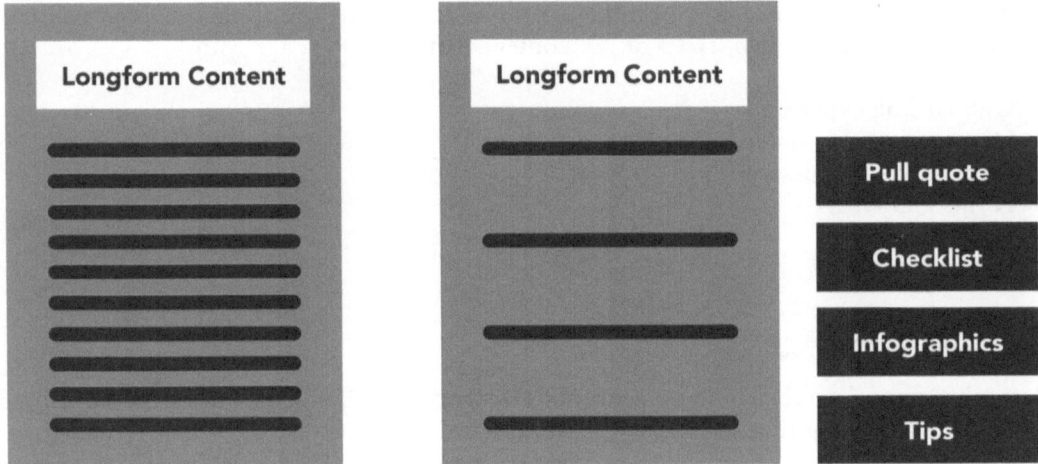

Figure 9.2 Creating microcontent from long-form content.

Not to be confused with microcopy, microcontent is another term associated with UX that needs a proper definition. While microcopy refers to texts and non-textual elements on an interface, microcontent is short-form content that helps users quickly digest traditionally long-form content. Long-form content is all around us. You can easily see it if you look up and scan your surroundings: books, articles, blogs, news reports, product descriptions, user instructions... and so much more! While long-form content serves the purpose of communicating information in depth, people need time to consume the content, which may not always work for users who are seeking immediate solutions to the task at hand.

This is not to say that UX writers should avoid long-form content. A lot of UX writing scenarios require long-form writing to meet compliance requirements (like end-user agreements) and user expectations (such as magazine articles or investigative reports). However, good UX writers need to know how to translate long-form content into microcontent for the purpose of different use situations, especially for mobile and screen-based reading (Figure 9.2).

Some quick tips for creating microcontent from long-form content include the following:

- Limit content to four to five chunks of information (each "chunk" should be digestible within 20–30 seconds).
- Use pull quotes to call out key "sound bites" that represent the gist of the content.
- Create visual content using combined graphics such as infographics or charts (including Venn diagrams and overlaying graphical data).
- Use carousel banners or cards to hold similar categories of information.
- Make a list (numbered for steps and bulleted for itemized objects) to allow readers to scan the content quickly—like this list!

Whenever appropriate, use short phrases rather than full sentences to minimize the number of words on the interface. It does take practice to master designing microcontent, but once you've gotten a hold of it, writing the UX genres in the following sections would be bliss.

Onboarding Experiences

In Chapter 1, we briefly talked about how UX writing is about the *whole* experience, not just isolated interactions with a physical product. If you're like the millions of smartphone users today, you probably do not remember the onboarding process when you're first setting up your new device. This is due to the meticulous amount of care and attention the UX writers of those products have given to the onboarding process. From the moment the user unboxes the new device to when they see the screen light up and then to the setup page where it says,

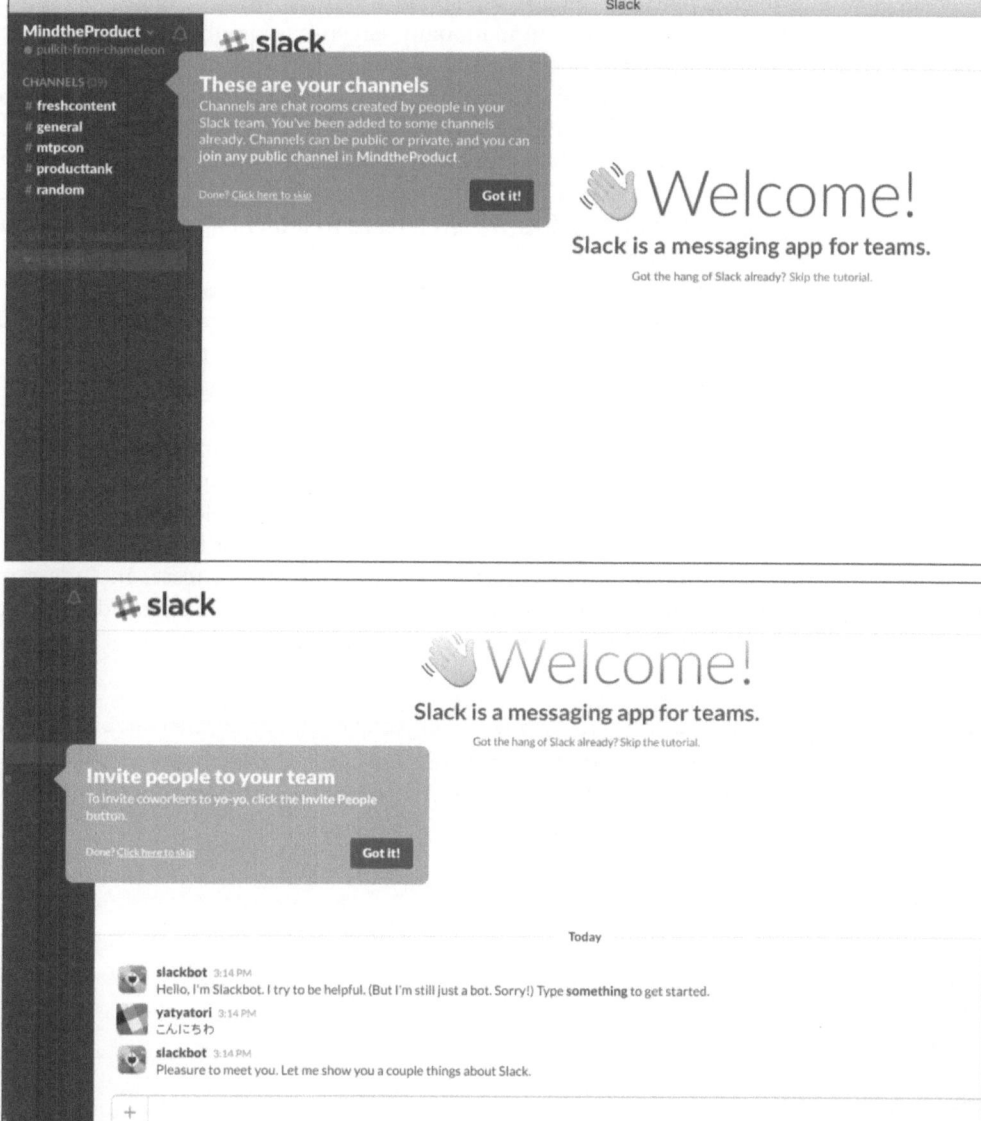

Figure 9.3 Walkthroughs can be an effective way to introduce users to basic features on a new platform.

"Let's Get Started," the user has already met a series of UX content. Every step of the way is meant to be encountered with ease and even excitement because good product designers know that getting started with a new device can be stressful, especially for novice users.

We present onboarding experiences as the first common UX writing task because it sets the tone for other UX writing situations by calling attention to holistic/whole UX writing and design. Onboarding is not limited to new product experiences, of course. Many digital platforms, such as emails, team management tools, social media, apps, and web management systems, require some form of onboarding training after product adoption. These situations require UX writers to pay attention to the common struggles, expectations, and requirements of users when encountering the platform. Figure 9.3 shows a typical "walkthrough" demonstration on a communication app that can be found in many screen-based onboarding experiences.

The best way to put users at ease is by writing in conversational mode. Imagine giving someone a tour of your home or school campus. Be inviting in your tone but not too formal, unless that is the brand persona you're striving to achieve. At any rate, assume that the users have minimal background in the nature of the platform—after all, they are new to it—but never talk down to them. A good way to gauge the best level of complexity in this interaction is by working with your product research team to understand the demographics of your users. As we have indicated in the first two parts of this book, understanding users is critical to successful UX writing, and it begins with empathy and research.

Non-textual elements in onboarding design should also be given careful treatment. Like the examples shown in Figure 9.3, use proper headings, body texts, and colors to distinguish the various components of a walkthrough. Importantly, give the users the option to skip unnecessary parts of the tour but allow them the opportunity to return to the demos in the future. As new features get added through product updates, include quick how-to demos along with the updates. As with the initial tour, keep those simple as well.

Last, in all onboarding experiences, time and progress are an invisible part of the process. If applicable, visualize the progress that the user has made throughout the setup so they can estimate the amount of effort needed to complete the onboarding process. Include a progress bar or the number of steps that the user has completed. This kind of consideration can give users a sense of assurance as well as an appreciation for the product.

Try this: **Pick a simple app** that you already use on your phone or another mobile device. **Create a list of questions that a new user may have** about the app based on your existing experience with it. Then, **write a few suggestions to address those questions** during the onboarding experience.

The app:

The questions:

The suggestions:

Help Guides and Contextual Tooltips

It should be every UX writer's mantra that every user query should lead to helpful answers. As you know, users may encounter new problems or questions about a platform after the onboarding experience. Writers play an important role in documenting and presenting answers for users in genres such as help guides and tool tips. Help guides are usually extensive instructional materials that are organized by issue type or task category. Contextual tool tips are microcopies that appear during the users' interaction with content to provide *in situ* information that helps the user perform desired tasks (see example in Figure 9.4).

In writing help guides and tooltips, it is important to begin with the most frequently asked questions (FAQs) so that high-level issues are taken care of in the first place. FAQs can be generated through tracked queries. Every time a ticket or a query is reported, you should ensure that it is documented properly with an organizer like a spreadsheet or table. This would allow auditing of these queries when you have received a sufficient amount of them to generate aggregated categories of issues. With these categories, you may then write help guides that respond to high-query problems. Alternatively, you may organize your help guides using the following grouping methods:

- Chronology of use
- Frequency of use
- Functional categories
- Expertise level (basic vs. advanced problems)

Keep in mind that every user's experience with a problem can be unique, even if they belong to the same type or category of tasks. Therefore, your help guides should offer options for users to search via multiple keywords (e.g., "pairing" vs. "connecting" vs.

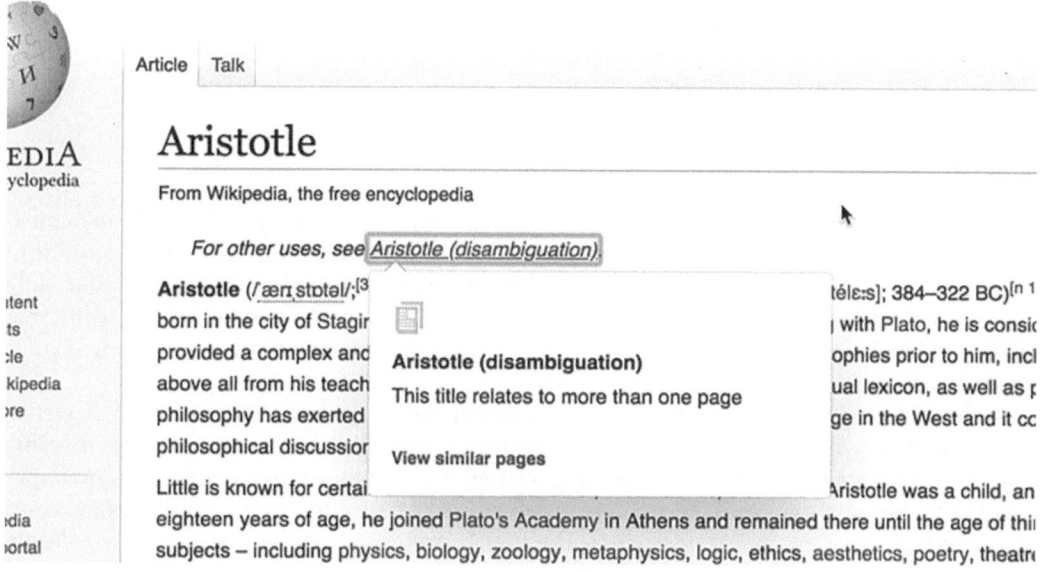

Figure 9.4 Pop-up tips to explain the inclusion of a separate entry page with a similar topic on Wikipedia.

"setting up" vs. "matching" for Bluetooth devices) and show recency via version information, release date, and other time-based details. This would help users identify the best solution to their problems and avoid using incorrect or outdated answers.

It would be smart to designate spaces for users to interact with each other in the form of short forums or votes (thumbs up or down). This interaction design serves two purposes. First, it creates a sense of agency (control) among the users and lets them support other people in the same community by offering tips and tricks for solving problems. If moderated well, this could lead to brand loyalty. Second, allowing users to provide feedback on the helpfulness of your help guides can give you directions for revisions or updates. You can also use this feedback to learn about the characteristics of your users.

As for contextual tooltips, be conscious of the amount of information given to users as they're performing a task. Since these tips pop up during an active interaction with the interface, they could hinder the task at hand and lead to frustration for the user. In Figure 9.4, you may see an example of a just-enough set of tips offered to the user when deciding which Wikipedia entry they prefer to reference. It does take up screen space, so be mindful of the pop-up design, especially on smaller or mobile devices.

As always, test your help guides and tooltips with actual users. Have a set of beta testers use your help guides to perform a certain task or resolve an issue with your product, then identify the way these testers interact with the guide. The same applies to tooltips for the purpose of usability and effectiveness. Finally, remember to devise a plan for updating help guides and tooltips based on the pattern of new releases for your product. For larger companies and services, a single-source approach to managing old and new guides can save you time and money. Refer to Chapter 7 for single-sourcing methods and strategies.

Error Messages

Inevitably, users commit errors when using products, even if they're familiar with the interface. Errors could lead to confusion and negative reactions that could cause the user to leave the product. UX writers are responsible for designing effective error messages that:

1 Inform the user that an error has occurred;
2 Offer quick solutions to recover from the error, and if appropriate;
3 Let the user submit an error report to the provider.

Good writers understand that visceral (involuntary, often emotional) reactions can be powerful influences on users' mental associations with your product. In error situations, users form visceral reactions quickly. Your error messages can help mitigate negative judgments or undesirable associations the user might make about your product. In turn, error messages that help the user recover from mistakes and continue their tasks might lead to positive user evaluations.

To design error messages that inform users about the error that has occurred, UX writers provide succinct, clear, and meaningful information about the error situation. This information should be written in human language rather than system language that is incomprehensible to users. If technical terms are involved, make sure to supply short explanations for the terms. Most errors occur because the user misunderstood the instructions or did not

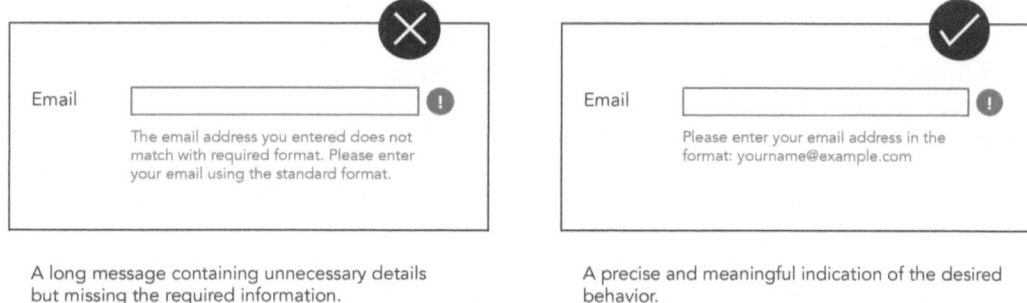

A long message containing unnecessary details but missing the required information.

A precise and meaningful indication of the desired behavior.

Figure 9.5 Short and understandable error messages help users learn what they need to do.

know what was required by the interface. Helpful error messages should contain examples or information about what the user is supposed to do (Figure 9.5).

Following the advice from one of Jakob Nielsen's (1994) 10 heuristics for interface design, error messages should apply noticeable visual treatments—such as color (like red), text weight (bold), and pop-up boxes—to increase user awareness about the error.

Never blame the user for committing the error. Use constructive and polite language to encourage the user to make another attempt at the task using the additional instructions provided in the error message. Creative director Sonia Gregory (2019) suggested that a good way to write courteous error messages is by reading them out loud—hearing how they sound and the possible tone the user might perceive from the writing. This can help avoid rude language that may be unintended but still perceptible to the user.

There are many more tips for designing good error messages than we could include in this chapter. So, we have compiled the following recommendations made by UX professionals around the world for your quick review:

- Avoid ambiguity: Don't tell the user, "An unnamed file with an unnamed object was detected."
- Instruct rather than insult: State, "Please enter your name," not "You didn't enter your name."
- Use neutral or positive words in messages: Say, "Please enter a valid zip code," not "This zip code is invalid."
- Be specific about the options to resolve an error: It doesn't help users learn the cause for the error if you simply tell them, "The item was either already moved or deleted, or access was denied."
- Embrace humor when possible. Consider this 404 error-page message: "Look! You've found a page that doesn't exist. Were you looking for _____? (link to an available page)"

Forms and Labels

As a writer who is used to crafting sentences and paragraphs, you probably haven't considered forms to be a form of UX writing (pun intended). Indeed, forms are often an easily overlooked element in content design, but they are everywhere. Citing the famously flawed voter ballot used in Florida during the 2000 U.S. presidential election, Jarrett and Gaffney (2009) put it aptly: bad forms can lead to serious consequences.

As you've learned in the previous chapters, a great deal of UX writing involves designing and organizing information so it can be used effectively by the target audience. Information architecture involves structuring content with organizational strategies such as forms and data sheets. These strategies help create groupings of content, guide users into providing relevant information, and help organizations create a database of customers and followers.

Although forms might look like an invention of the modern age, their use has actually predated computers and even papers. At a time when the writing system (think alphabets) was still being invented, early civilizations, including the Egyptians, were already using form-like documentation methods to mark their collections of crops and goods. They created topical categories carved into tablets to note seasonal harvests and transactions or sales. At any rate, forms are a human way of accounting. They afford a quick and easy way to categorize and document information.

UX writers are expected to be experts in creating forms. UX writers work with form design to help their organization or business curate and organize user information. Within forms, labels are short headings that indicate what users need to do or enter for a particular field in a form. Labels can be presented as descriptive phrases, like "Enter password," or as questions, like "What's your address?" It takes practice to become proficient at designing user-friendly labels on forms. Here are some tips to begin with:

No one likes filling out forms. The practice is often deemed tedious or "busy work" that takes time away from other more desirable activities. So, writers should aim to make filling out forms effortless. The best way to do so is by using autofill. Apply browser cookies and cache to enable forms to find routine information such as the user's first and last name, primary email, and home or shipping address. By taking care of these uninteresting fields, the user may then focus on other parts of the form that are more meaningful and important to them.

Brevity is well appreciated in all forms. Do not ask users to provide unnecessary or redundant information (like a home address *and* shipping address). Give users the option to provide additional information only if they want to. One strategy to design short forms is to collapse optional fields. Users who want their purchased items shipped to a location other than their home address may check the optional box to provide a separate mailing address.

Provide effective error messages. Users frequently commit mistakes when filling out forms because this is not a natural human activity. Good UX writing, as we have indicated in the sections above, includes providing users with succinct instructions to recover from errors when they occur. Error messages need to be meaningful (tell the user what happened) and effective (how to recover quickly). Of course, it is even better to design forms that mitigate common errors altogether. One way to do this is by writing clear and concise field labels that let users know what they are required to do and what's optional. Visual cues or indicators such as red asterisks (see Figure 9.6) are commonly used to help users identify requirements easily.

Remember to design forms for all kinds of interfaces. Mobile and small-screen devices can be tricky to work with given the lack of digital real estate to display forms appropriately. The most helpful tip to design for limited screen space is by keeping forms to a single column. Keeping in mind the advice to maintain brevity, limit forms to three to five fields per page. If you have additional fields that won't fit on the screen, employ a multi-step form design rather than a long form that makes the user scroll for what might feel like indefinitely.

Figure 9.6 Use visual indicators to show users required vs. optional inputs.

Legal Notices

Because of their linguistic proficiency, UX writers are often asked to collaborate with legal specialists to design end-user agreements and legal notices for compliance purposes. Common legal notices include:

- Terms and conditions
- License agreements
- Copyright notice
- Disclaimers and limits of liability
- Privacy and accessibility notice
- Governing law

These notices are usually required by law to be presented to end users so they can make informed decisions about their usage and engagement with a technology or platform. However, you may have already realized that these contractual languages can be complicated, lengthy, and boring, and hence users usually ignore them. In fact, a study done by Kevin Litman-Navarro for the *New York Times* in 2019 revealed that the vast majority of privacy policies at 150 tech providers exceed the college reading level.

Another study by communication professors Jonathan Obar and Anne Oeldorf-Hirsch (Berreby, 2017) showed that only a quarter of their 543 test participants bothered to look at the fine print before clicking "agree to our terms and conditions" to join a made-up social network. The other three quarters had supposedly agreed to give the network their future firstborn children, per paragraph 2.3.1 of the terms of service. All jokes aside, it would be considered unethical to cast blame on the users rather than assuming responsibility to present important legal notices in an accessible manner. UX writers have the obligation to help their organizations design better user experiences with legal notices.

To design truly usable and welcoming legal notices, UX writers can apply information grouping strategies to reduce reading fatigue. The psychology of cognitive processing tells us that chunking large content into smaller categories can give people the illusion that less effort is required to process information. As in Figure 9.7, writers can use block modules, header sections, short descriptions, and color contrasts to organize smaller units of terms and conditions.

In the same figure, you may also notice the effective use of a table of contents on the left panel to create a scannable outline of the full legal notice.

UX writers can help users quickly skim for important parts of the notices they wish to read by summarizing major services and implications, like in the example below:

Table 9.1 A description of service (left) summarized into two sentences (right).

Original description of service	*Summary of service*
The500, Inc. ("The500," "we," "us," or "our") provides a platform via its website, messenger service, and apps (the "Site") to a community of registered users ("users" or "you") to engage in a variety of activities, including to upload and display photographs and video ("Visual Content"), share comments, opinions, and ideas, promote Visual Content collections, participate in contests and promotions, register for premium membership accounts ("Premium Accounts"), and license Visual Content to other users through our distributors (individually or collectively, the "Services"). The foregoing list of services is not all-inclusive, and additional services may be offered by us from time to time. The following are the terms of use ("Terms") for using the site and the services.	We develop a photo community, provide services to create online portfolios, and license photos through distributors. We will develop more features and services in the future.

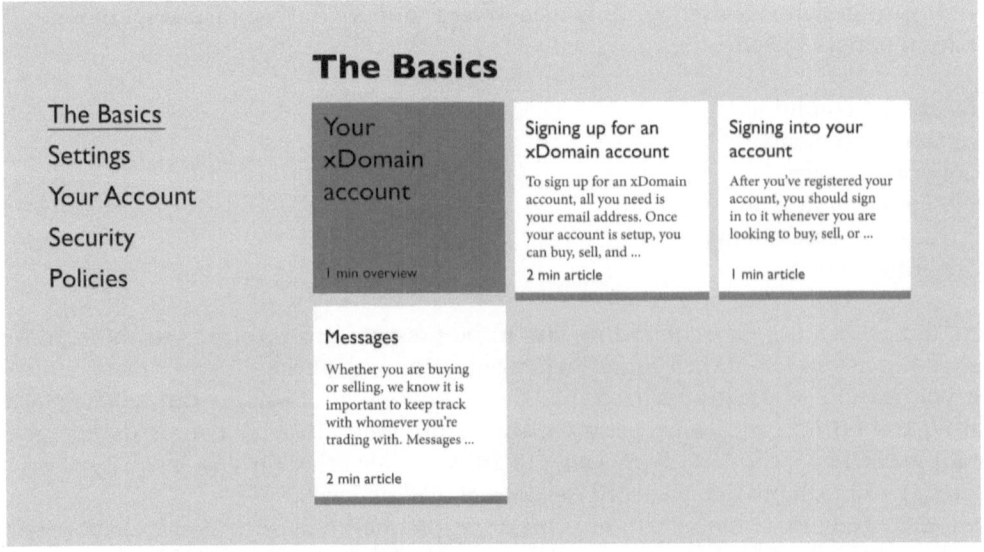

Figure 9.7 Use visual and textual chunking methods to create smaller reading tasks for users.

By now, you should already know that a conversational tone can serve to make technical content more personable to users. UX writers can help edit legal notices and design a tone of voice that speaks *with* rather than *at* users. For example:

Instead of saying: This service does not collect or share personal information. It is our privacy policy not to collect cookies, local storage, or browser/device fingerprinting for marketing purposes.

Say: We do not track you or your private information.

Instead of saying: You may terminate your agreement or close your account with RapidCart at any time, effective the last day of your subscription term, by providing a written notice to support@rapidcart.com indicating your termination of the relationship.

Say: We'd hate to see you go. But if you want to leave, you may terminate your account at any time by sending us an email at support@rapidcart.com.

Now, try it yourself. Translate the following language into a more conversational tone.

Original:

"This Agreement shall be effective as of the date (the "Effective Date") the User accepts the terms herein or first accesses, downloads or uses any of the services or information (collectively, the "Services") on the site and shall remain in effect for so long as the User uses or accesses any of the Services (the "Term"). Upon termination of the Term, the User shall no longer be permitted to use or access the Services."

Conversational:

_____

_____

_____

_____

Finally, to cultivate a culture of paying attention to corporate and technological legality, UX writers can help educate users about the importance of reading legal notices. Include in the notices why it is important for them to understand the terms of use before agreeing to abide by them. Provide options to learn more about the technical language and personalize their privacy settings. To give users more opportunities to learn how certain services affect their personal use of a technology, create channels that allow users to return to the legal notices even after accepting them. This may provide users a chance to understand the stakes of their engagement with a particular platform.

Settings and Specs

Users need information about products in order to use them to accomplish tasks efficiently. UX writers are the translators who convey system information to everyday users about how they can do great things with the product. This last most popular kind of UX writing, made almost entirely of texts, is known as settings and specifications, or specs. Do not underestimate the power of good UX writing by producing settings and specs. Basecamp founder and CEO Jason Fried (2013) put it aptly in a tweet:

> "Here's what our product can do" and "Here's what you can do with our product" sound similar, but they are completely different approaches.
>
> (n.p.)

And these two approaches yield different experiences on the part of the user. Designing user-centered settings information requires careful consideration of user needs as well as use contexts. The definition phase of design thinking described in Chapters 3 and 5 provides excellent research methods for you to understand these requirements.

Start by identifying the benchmark for default settings. Default settings are values pre-set for the user. They define desirable application behaviors for the product user. Figuring out the defaults is probably the most challenging job in this writing process. Thankfully, there are established ways to approach benchmarking. If your product is similar to other market-place products, you can first review and compare its product property with those of similar products in your industry. Then, analyze common design features in those products' settings to find an initial benchmark. Apply that benchmark to your product and test it with a selected sample of users—novice and power users—with a series of tasks to determine gaps in the initial benchmark. With the findings from the tests, remove unnecessary features and generate additional ones customized to your product. Set that as the beta default setting and test it again with users in different use contexts. Iterating this process will help you achieve the desired level of settings that best serves your users.

Settings should welcome users to learn about the different aspects of your product and allow them to make personalized adjustments. It should not scare them away from toggling the configurations. Showing users the ease of personalization can increase their level of comfort with your product and lead to greater satisfaction. UX writers pay close attention to the mental models and references users apply to make sense of their products. Follow the design methods presented in Chapter 6, such as card sorting and affinity diagramming, to determine the best architecture for your product settings. Help users reduce cognitive loads by placing complex features under *advanced* or *other* settings.

Specs serve to inform users about the functionality of a product so they can learn to adjust the appropriate settings. UX writers recognize a difference between specs and user help guides. While both share some similar goals of supporting users, specs tell users how something works rather than troubleshooting. When designing specs, apply a similar approach as writing legal notices that we covered earlier—use chunking and a table of contents to create scannable documents so users can find what they are looking for quickly. Remember that users are unlikely to be system experts, so be sure to minimize jargon or technical names in the specs. If you are required to use the formal names for parts of your product, supply a common name or explanation for each of the proper names.

UX writers are cognizant of user diversity and cultural differences. It goes without saying that designing for adaptability in terms of users' expectations is going to make your specs more globally acceptable. If your product is marketed internationally, be sure to include precise units of measurement in the specs. To prevent mistakes in translation, avoid cultural expressions, abstract fashionable terms, compound sentences, and unique acronyms.

Designing Content with Style and Tone Guides

Whether you are creating an onboarding training program, screen messages, forms, or product specs, it is important to adhere to your organization's style guide and tone guide. A style guide specifies the aesthetic rules of all user-facing content, including the use of colors, typography, layout, and other visual elements. A style guide's purpose is to help you create uniform content that establishes a recognizable visual identity. A tone guide indicates the writing style for your content, including mood, voice, and technical aspects of writing such as spelling and punctuation. Like a style guide, the tone guide helps you in creating content that *sounds* consistent across your products.

Think of your content as having a persona of its own. For public users, your content should reflect an identifiable character with a unique personality. Style and tone guides assist you in achieving that brand identity. As an example, consider Mailchimp's (an email marketing service) identity guidelines:

> Structural elements like our logo, color palette, and typography keep us grounded and consistent. These core components work together to ensure our brand is recognizable wherever it appears.
>
> The flexible elements of our brand celebrate creative expression. Our new approaches to illustration, animation, and photography allow us to communicate with a wider range of emotions, take more risks, and showcase the creativity of our artists and designers.
>
> (Source: https://mailchimp.com/design)

These specific ways of presenting a brand can help customers and users recognize the brand when using its content, even when the brand's logo or wordmark is absent. These guides also help users relate to the brand through its personable content, thus enhancing the overall UX.

When users recognize and remember a brand's identity, they are more likely to stay loyal to the brand and even recommend it to others. Consider another example: Starbucks. You know you're in a Starbucks coffee shop when you are in one—the layout, the colors, and the menu display are always consistent. They even have a uniform "vibe" in these shops, so much so that "Starbucks music" is a genre on its own (and you can download their playlist from the Starbucks app). As a UX writer, you play a critical role in designing the identity guidelines for your organization and specifying the style and tone of your content.

Designing Non-Textual User Interfaces: Video and Voice

Beyond texts and static graphics, UX writers also work with videos and voice interfaces in content design. While most of the UX content types covered in this chapter pertain to text-based writing, we expect new practices in virtual and immersive environments that will require new skills for designing content in non-textual ways. Here we offer some key recommendations for video and voice design.

Video UX has been extensively studied in instructional design and digital entertainment settings. Videos give users a different way to engage with content by combining sounds, movements or animations, and transitions. They depict meaning in ways that texts alone could not. However, videos aren't always superior to texts. They contain limitations in terms of formatting and user control. We highlight these two concerns in the discussion here.

In designing a good UX for video content, make sure first that the video format is accessible to all users. Accessibility includes the device's capacity to display the video appropriately and assistive elements that support various user needs. Since videos require greater bandwidth to be hosted and run on screens, writers should consider how their users would access the video content (e.g., on a desktop vs. on a phone) and determine the format of the video design. To ensure accessibility for users with accommodation needs, include features such as closed captioning, speed control, and transcripts. These features give users better control over the ways they prefer to engage with the video content.

Another tip for UX writers is to pay attention to video length and the scripting of the content. Most users today prefer shorter videos for tutorials and instructional purposes and lengthier videos for entertainment. For videos to achieve optimal UX, writers should know what users aim to get from videos and design them according to their contextual needs. For example, if the video serves to teach users how to perform a task (like cooking videos), it should include overviews, outlines, and summaries at the end that help users learn the task effectively.

Similar to videos, voice interfaces have design challenges that need to be attended to by UX writers. The voice user experience has been focused on voice command systems that supplement visual and textual interactions between users and devices. Voice interfaces afford a "hands-free" kind of engagement with content but contain usability issues such as accuracy, flexibility, and social acceptability.

Whether for phones, TVs, or car consoles, voice controls need to ensure accurate delivery of the information being requested. Users typically use voice commands to search for information (like, "How long would it take to get from home to _____ [insert place]?") or perform simple tasks (set an alarm, find movies, track fitness activities). Working with programmers, UX writers play a role in using natural language rather than technical/system language to achieve accuracy in information output (Figure 9.8). By devising and implementing a way for aggregating and analyzing user voice input, writers can create heuristics that guide the voice interface in understanding user queries and subsequently providing accurate feedback.

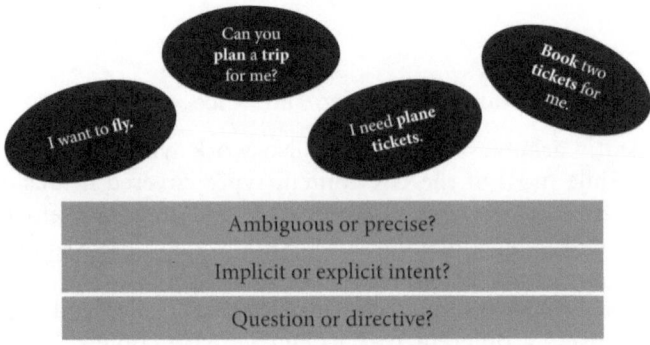

Figure 9.8 Voice interfaces must consider non-linearity in user input to understand queries.

Good voice UX requires flexibility—offering multiple commands to accomplish the same task—to accommodate different user speech behaviors. To design for universal use, writers work with machine learning designers to train voice AI to pick up various speech patterns and accents. Over time, voice agents will become better at understanding commands and adapting to different human conversational modes. Voice interfaces would be more widely accepted and used if users found it effortless to interact with them.

To leverage the affordances of voice interaction, UX writers still have some work to do to increase its social acceptability. Social acceptability is a heuristic measure for how well voice interfaces integrate with the kinds of environments in which users would interact with voice agents. Current voice interfaces still lack acceptance by users in public settings, such as their workplace or when they are not at home. Based on a survey of 500 consumers in the U.S., the public research agency Statista reported that 51% of the users would use smartphone-based voice assistants in the car, 39% at home, 6% in public, and only 1% at work (Baker, 2018). To enhance voice UX beyond personal spaces, writers need to consider social cues and cultural conditions surrounding users and design interactions suitable for these contexts.

Conclusion

UX writing involves a lot of tasks and manifests in many different genres. In this chapter, we have highlighted the six most popular types of UX writing that you should get acquainted with. They are onboarding experiences, help guides and tooltips, error messages, forms and labels, legal notices, and product settings and specs. As a writer, you will very likely encounter these types of content in your professional practice, regardless of the industry you serve. As well, you will be asked to work with content delivery beyond textual modalities, such as videos and voice. No matter which of these UX content you create, be sure to maintain a user-centered design mindset and practice design thinking strategies.

Chapter Checklist

This chapter has introduced you to many tricks and tips for designing user-centered content across six popular UX writing genres. To facilitate good UX in these content types, we recommend practicing these strategies in a variety of contexts where these UX writing genres may manifest.

- *Understand how microcopy is different from microcontent.*

  - Microcopy refers to individual elements on a user interface.
  - Microcontent includes short-form components created from long-form content.

- *Focus on the whole user experience through onboarding design.*

  - Strive for an inviting tone to ease users into new environments.
  - Use graphical/visual elements to help users distinguish between components of a walkthrough demo.
  - Visualize the progress users have made in the onboarding process.

- *Support users with effective help guides and contextual tool tips.*

  - Track user queries to identify the most FAQs.

- Allow users to interact with one another through community forums.
- Design tooltips that are contextual and non-intrusive to the user's activity.

- *Provide meaningful error messages to help users recover from mistakes.*

  - Inform users that something wrong has occurred.
  - Provide quick and effective ways to resolve the issue.
  - Use polite and appropriate language to encourage users to make new attempts.

- *Apply information architecture strategies to create forms that make sense to users.*

  - Only ask for necessary information; eliminate fluff from forms.
  - Keep forms short.
  - Design for all kinds of interfaces.

- *Make legal notices more conversational and readable.*

  - Use chunking to split legal notices into manageable, smaller sections.
  - Talk with users, not at them.
  - Educate users about the importance of reading legal notices.

- *Teach users how to optimize their experience through product settings and specs.*

  - Do marketplace research to identify a benchmark for default settings.
  - Design specs that are adaptable to cultural diversity.

- *Design with style and tone guides.*

  - Use style guides to determine the visual identity of your brand.
  - Use tone guides to create a consistent voice in your content.

- *Design videos and voice interfaces that support user needs.*

  - Give users a sense of control over video content.
  - Develop voice interfaces to be accurate, flexible, and socially acceptable.

Discussion Questions

1 In addition to the examples given in this chapter, come up with two to three more kinds of microcontent to show their difference from microcopy.
2 How can visualization help enhance UX in onboarding design?
3 What are the major challenges in designing help guides, error messages, forms, and legal notices? What smart ways have you personally experienced that make the design of these genres more user-centered?
4 How do videos and voice interfaces differ from print or screen-based content? What do you need to pay attention to when designing non-textual content?

Learning Activity

Using Figure 9.1 as a template, create a collage of UX writing genres with actual samples. Create a document or slideshow with the various categories using screenshots, photographs, or other visualization methods (Figure 9.9).

Then, compare your "genre board" with a friend or another student and see what similarities and differences you've observed. Discuss what helped you select the specific genre and how it stood out for you as a UX writing product.

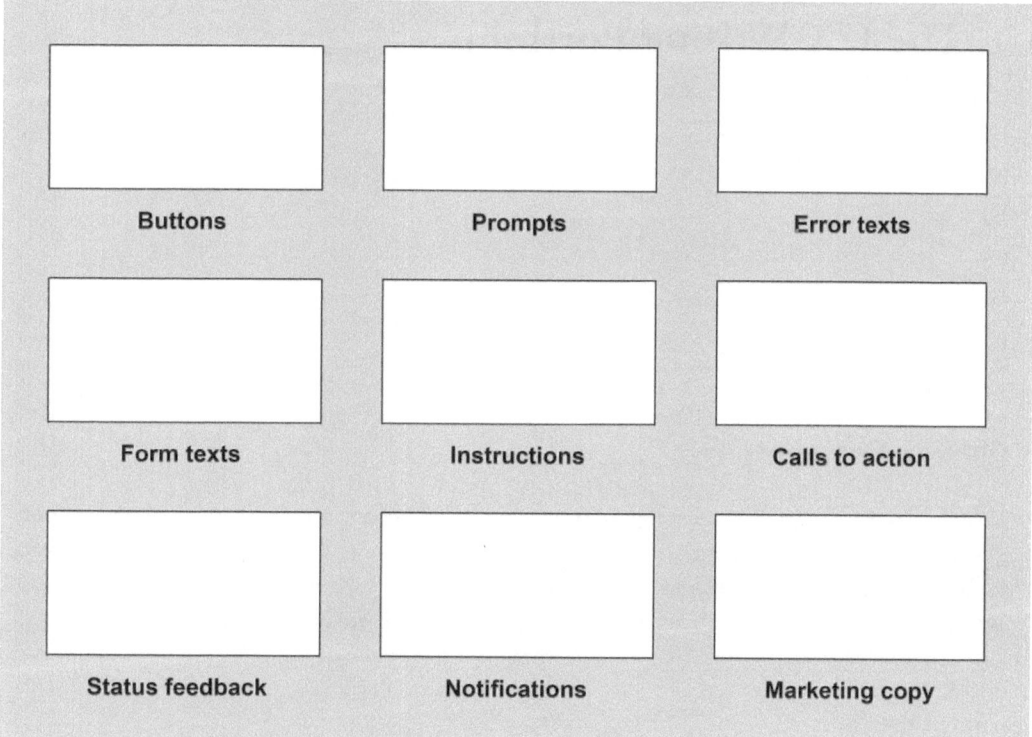

Figure 9.9 Blank template for popular UX writing categories.

References

Baker, J. (2018). Voice user interfaces (VUI)—the ultimate designer's guide. Muzli-Design Inspiration. https://medium.muz.li/voice-user-interfaces-vui-the-ultimate-designers-guide-8756cb2578a1

Berreby, D. (2017, March 3). Click to agree with what? No one reads terms of service, studies confirm. *The Guardian.* https://www.theguardian.com/technology/2017/mar/03/terms-of-service-online-contracts-fine-print

Fried, J. [@jasonfried]. (2013, November 13). *"Here's what our product can do" and "Here's what you can do with our product" sound similar, but they are completely different approaches.* [Tweet]. Twitter. https://twitter.com/jasonfried/status/400733165964099584?ref_src=twsrc%5Etfw

Gregory, S. (2019). Best error messages: 5 tips for a user-friendly experience. Freshsparks. https://freshsparks.com/user-experience-tips-best-error-messages/

Jarrett, C. & Gaffney, G. (2009). *Forms that work: Designing web forms for usability.* Morgan Kaufman.

Litman-Navarro, K. (2019, June 12). We read 150 privacy policies. They were an incomprehensible disaster. *The New York Times.* https://www.nytimes.com/interactive/2019/06/12/opinion/facebook-google-privacy-policies.html?mtrref=undefined&assetType=REGIWALL&mtrref=uxdesign.cc&assetType=PAYWALL&mtrref=www.nytimes.com&gwh=E110DBE7D76868C83A851BAAF1835998&gwt=pay&assetType=PAYWALL

Nielsen, J. (1994). 10 usability heuristics for user interface design. Nielsen Norman Group. https://www.nngroup.com/articles/ten-usability-heuristics/

10 The UX Writing Portfolio

Chapter Overview

This chapter covers tips for creating and presenting a UX writing portfolio. You will learn about the rationale behind keeping a portfolio and the key components reviewers generally expect to find in a portfolio. We discuss three major sections of a UX writing portfolio and provide examples and resources for building your own portfolio with ease and confidence.

Learning Objectives

- Describe the purpose of a portfolio.
- Distinguish between working vs. presentational portfolios, static vs. dynamic portfolios, and social showcasing platforms vs. personal websites.
- Select key components to include in a UX writing portfolio.
- Set up and gather peer review feedback on your portfolio.

Why Do You Need a Portfolio?

Real World Snapshot 10.1: A UX Portfolio Situation

Ren has just graduated from a technical communication degree program, and he is excited to find several job openings in UX writing. He is preparing to apply to some of them, but when studying the job ads more closely, he was surprised to find that they all required a "UX portfolio" along with the application package. Having only kept a simple resume all throughout his college career, Ren is a little panicked about this portfolio requirement. He doesn't even know what a portfolio should look like.

Is it like a collection of drawings? Photographs? Screenshots?
Is it like a website? A blog?
Is it like a suitcase full of things I have made in the past?
Is it like a binder of writing samples?

DOI: 10.4324/9781003274414-13

Thankfully, Ren has a friend, Rick, who has worked as a hiring manager in a creative agency, and Rick recommended this chapter as a starting point for Ren to build his UX portfolio.

If you recall, in Chapter 2, we touched on an attribute of the design thinking mindset that is "Show, don't tell." Almost every UX position now requires its candidates to provide documented experience to *show* that they have the necessary skills and traits to succeed in the position. A portfolio serves this purpose. It is a dossier, a collection of selected artifacts about yourself that showcases your strengths and potential. When designed and presented well, a portfolio can *demonstrate* your creativity and problem-solving abilities, which are desirable in all UX writing jobs. If you will, your portfolio is your spokesperson, your brand identity, and your sales pitch when you're not there physically to make a case for yourself. This chapter aims to highlight the key aspects of portfolio development for UX writing professionals and guide you through the process of creating a persuasive portfolio.

Where do portfolios come from? While the history of portfolios in the design profession may be blurry, we could trace their emergence to the important influence of the advertising industry in the 20th century. Copywriting and graphic design jobs were in high demand from the 1960s to the 1980s, when many creative minds aspired to be the next David Ogilvy (of Volkswagen fame) and Leo Burnett (creator of the Jolly Green Giant, Pillsbury Doughboy, and Tony the Tiger). The Netflix show *Mad Men* was created to capture this golden time of advertising. In the show, you could see well-suited writers and designers using hand-carried cases to keep their design samples (Figure 10.1). The idea is to give reviewers (or recruiters, hiring managers, etc.) a quick reference to the writer/designer's work during meetings or interviews.

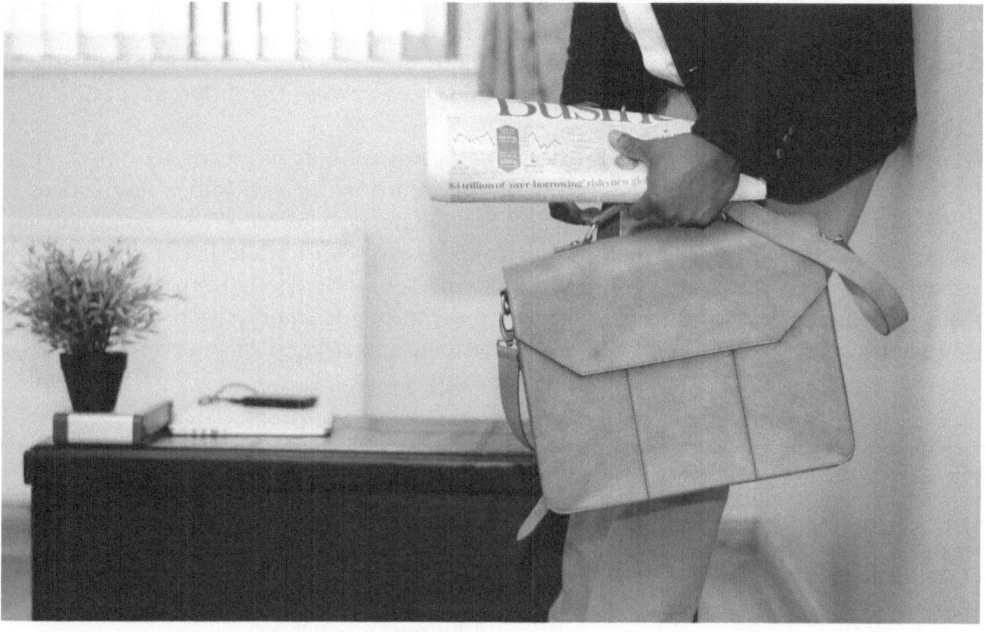

Figure 10.1 Physical portfolios are the origin of digital portfolios.

Source: Photo Adeolu Eletu on Unsplash.

Today, portfolios—portable folios—continue to be used in hiring situations across many creative industries, including UX. There is an obvious shift from physical portfolios to electronic or digital portfolios given the rise of the web. Writers and designers publish their work online to make them more available to reviewers. What's more, the internet also affords greater interactivity for presenting sample work, allowing the creator to include richer media (like videos, links, and animated displays) to enhance the reviewing experience.

Ultimately, you need a portfolio to get a job. Even if you aren't in the market for a new job, a portfolio documents your development as a practitioner and can be useful in other scenarios, such as:

- Winning a new project account (especially if you work freelance).
- Demonstrating your credibility as a professional (useful when you're applying for innovation grants and funding).
- Chronicles your growth as a practitioner (like a resume but much more expansive).
- Providing examples and case studies to other aspiring professionals (like students) so they can learn from your experience.

Where Should You Begin?

As in any creative genre, a portfolio doesn't really have set standards, but you'll know one when you see one. That's not very helpful, is it? Based on our survey of online portfolios and the advice that seasoned experts have given on different channels, we have created the following comparisons to help you make decisions about creating your own portfolio:

- Working portfolio vs. presentational portfolio.
- Static portfolio vs. dynamic portfolio.
- Social showcasing platform vs. personal website.

The first consideration in creating a portfolio is to distinguish a working portfolio from a presentational portfolio. Below is a comparison table showing the different qualities of the two (Table 10.1).

A working portfolio is a personal documentation tool. It is meant to be an ongoing collection of your work (artifacts) and to be viewed only by you and a few others from whom you may be gathering feedback from. For the purpose of record-keeping, you should create and maintain only one working portfolio so you can update it easily and not have to worry about version control. In contrast, a presentational portfolio is a public document. It is meant to be used during your job application and interviews to supplement your other materials. It is a polished *showpiece* that contains several of your carefully selected

Table 10.1 Working portfolio vs. presentational portfolio

Working portfolio	*Presentational portfolio*
Private	Public
Accessible to you and a selected few	Seen by employers
Contains all your artifacts	Contains selected artifacts
Only one	Can have several versions
Work in progress	A finished product

Source: Williams (2009).

artifacts—taken from your working portfolio—to demonstrate the skills and experience that the employer is looking for. Because of this selectiveness, you may create several versions of your presentational portfolio to appeal to different audiences and situations. For example, you may create a version of your presentational portfolio that contains some copy you've written for a mobile app if you're applying to a UX writing position at a social media company. However, if you're applying for a job that focuses on social media research at a UX agency, you may want to choose artifacts that reflect your user research and testing experience. Certain employers would even limit the number of artifacts in your portfolio, so it is important to remain flexible in crafting a presentational portfolio and follow directions.

Next, you should consider whether you need a static portfolio or a dynamic portfolio. The difference between the two is the level of interactivity they each afford. A static portfolio is usually brief and focused and requires minimal interaction by the reviewer with your materials. The objective of a static portfolio is to communicate your expertise quickly and directly. Static portfolios can be built with common applications like word processors (MS Word, Google Doc) and slideware (MS PowerPoint, Google Slides) and saved as PDFs. These portfolios can be viewed offline and even printed out by the reviewer if so desired. Figure 10.2 shows such an example.

A dynamic portfolio, in comparison, requires the reviewer to engage more intensely with the materials to *experience* your content. Dynamic portfolios typically feature digital

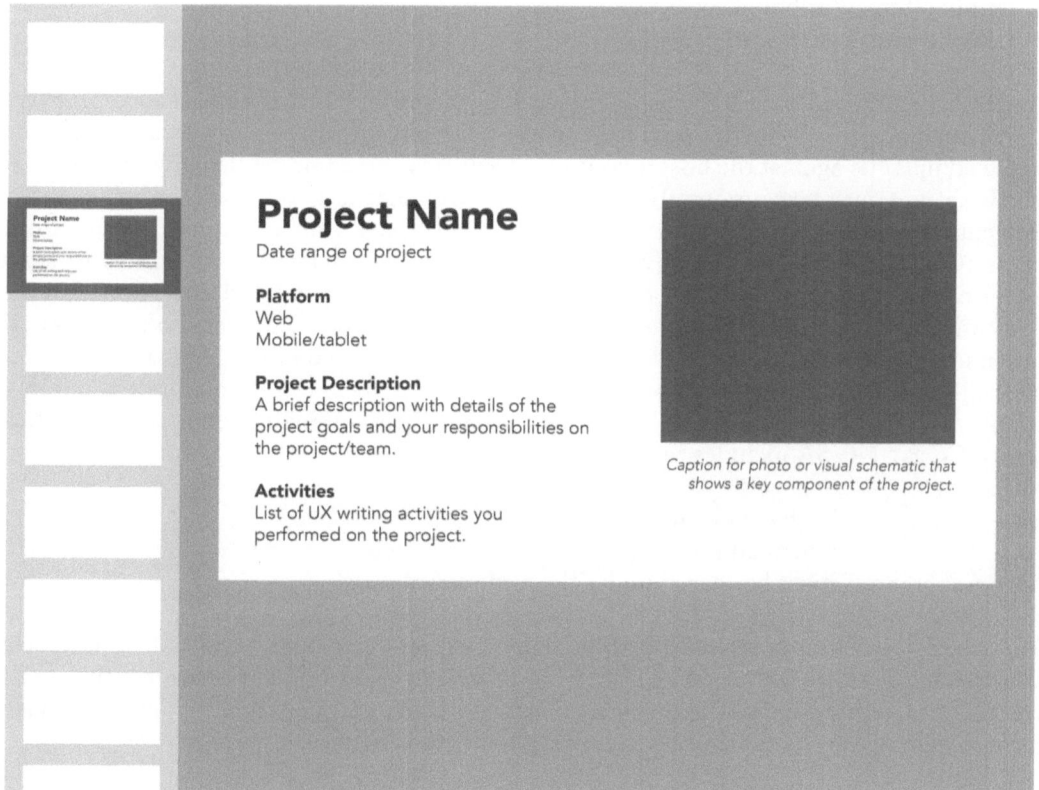

Figure 10.2 A static portfolio presented as a slide deck.

components such as clickable links, expandable images, videos, and interactive prototypes that allow the viewer to interact with your ideas. However, dynamic content takes more time to build as well as consume. As with most decisions you make about content design, you should consider the context in which your portfolio is required and how your audience (reviewer/hiring manager) may prefer to interact with it.

The last question to tackle is *where* to publish your portfolio. We have mentioned that most portfolios live online today due to accessibility and modality/interactivity preferences. Yet, there are two main options for you to choose as the host of your portfolio: a social showcasing platform and your personal website.

Due to the rising number of creative professionals across industries, as well as the social networking needs of these professionals, many platforms have been created to let individuals share their work openly to gather reviews and give comments to others. These social showcasing platforms allow users to create an account and own a dedicated page/space to curate their work samples. Being on these platforms increases the users' chance of getting discovered by recruiters. As well, since they are built to be social networking sites, users can respond to each other's portfolios and create a supportive community for themselves. Below are some examples of popular social showcasing platforms:

- Adobe Portfolio (https://portfolio.adobe.com)
- Behance (https://www.behance.net)
- Cargo (https://cargo.site)
- Dribble (https://dribbble.com)
- UXfolio (https://uxfol.io)

An important consideration for using these platforms is intellectual and creative property. Under the fine print of the user agreements on these platforms, you sign away your rights to file complaints against the host if your work (idea) is copied or "stolen" by another user. Likewise, you must be cautious about copyright infringement if your work sample contains elements that aren't your own. A good practice is to share your work via Creative Commons (CC) licenses (https://creativecommons.org) to give permission for others to share and use your creative work under the conditions you select with a CC license. For example, you can assign a CC BY-NC-SA license to your creative work. This license allows users to distribute, remix, adapt, and build upon the material in any medium or format for noncommercial purposes only, and only so long as attribution is given to the creator. If you remix, adapt, or build upon the material, you must license the modified material under identical terms. CC BY-NC-SA includes the following elements:

BY – Credit must be given to the creator
NC – Only noncommercial uses of the work are permitted
SA – Adaptations must be shared under the same terms

The second option to host your portfolio is by creating a personal website. Just like social showcasing platforms, there are many website builders and domain providers that make creating a dedicated webspace quick and easy. To start, you may find a content management system (CMS) like one of the following to acquire an account:

- WordPress (https://wordpress.com)
- Squarespace (https://www.squarespace.com)

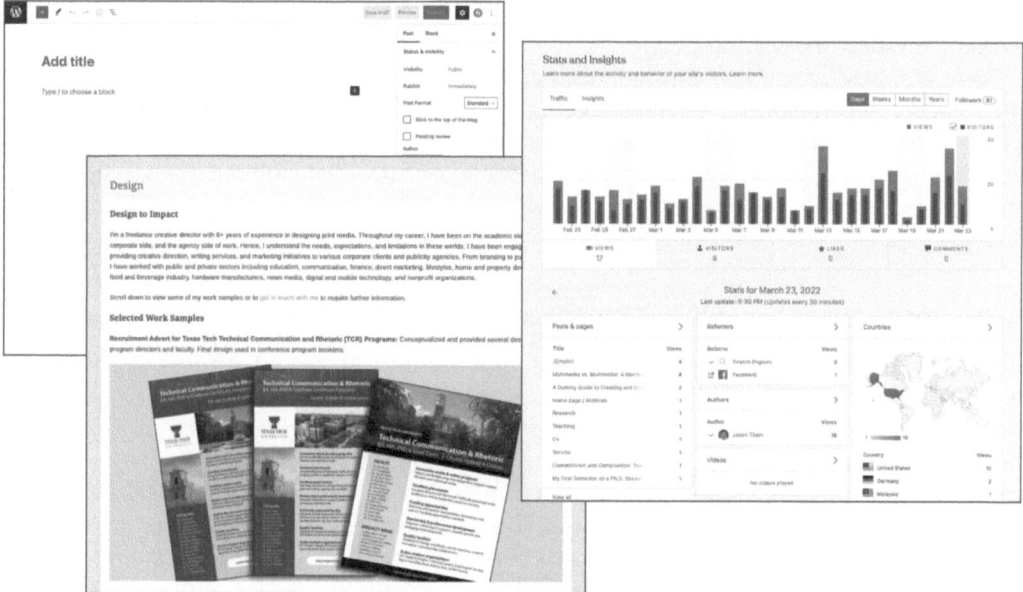

Figure 10.3 A collage of WordPress pages showing the front-end display and back-end dashboard of a portfolio. Left to right: the black composing page, the front-end view, and site analytics powered by the platform.

- Wix (https://www.wix.com)
- Google Sites (https://sites.google.com)

These CMSs do not require any coding/scripting (HTML, CSS) knowledge, although knowing some of the languages would help you personalize your site. There is also an abundance of ready-to-use templates suitable for portfolio display purposes on these CMSs. Figure 10.3 shows a collage of the front and backend of a WordPress CMS used by one of us (Jason) to host his portfolio.

Notice that you may also gather web analytics (traffic, hits, and activities) to learn about the popularity of your portfolio through these insights. They might be useful if you use the trends to optimize the content and publishing routine of your portfolio (like the best day of the week to post an update). Using a personal website to host your portfolio frees you from the constraints of structure and organization on social showcasing platforms, but a personal site would require more effort on your part to create interactions with your audience and peers.

What Does a UX Writing Portfolio Look Like?

Back to that initial question: So, what does a UX portfolio really *look like*? Everyone has slightly different assumptions about what a portfolio should look like. Even hiring managers who review thousands of portfolios may still be surprised by the diversity in portfolio presentation and design. The popular UX writing online collective led by Yuval Keshtcher has curated a yearly list of "Best UX Writing Portfolios," and it shows all kinds of organization, project narratives, and personalities (for example: https://yuvalkeshtcher. medium.com/best-ux-writing-portfolios-2019-update-7e6a066631af).

There are, however, a few common expectations. To achieve the purposes of a portfolio stated earlier—to introduce yourself, to highlight your expertise, to showcase your projects, and to invite connections—a portfolio needs to include three main components: (1) information about yourself, (2) descriptions about your problem-solving approach, and (3) actual project samples and results. In the following sections, we discuss these components in detail.

Component 1: About Yourself and Your Goals

Your portfolio is a professional reflection of your identity. It should impress upon viewers the kind of person, style, and character you would be on the job. Reviewers of portfolios often consider not only the qualifications of a candidate but also how well of a colleague this potential candidate might make. So, your portfolio should first and foremost communicate your personality and values, which would make you a desirable professional.

Start with a personal introduction about yourself. Be authentic and real about your personality, but don't try to fabricate a forged persona here. This is a space for you to share any unique or impactful experiences you had that brought you to UX writing. Share why you are passionate about this profession and what your goals are. Some good questions to guide your writing here are:

- What am I really good at?
- Which UX writing activities do I really like to do?
- What differentiates me from other designers?

Give your readers just enough details but do not overwhelm them (you will know this if you do a peer review; more on that later). A good guideline is to highlight the following:

- Your name and pronouns
- Direct contact information (phone and email)
- Social handles (LinkedIn, Twitter, Facebook, or social portfolio accounts)
- A quick narrative about your passion and goals

Keep in mind that the introduction section should only be a point of attraction, not the real juice of the portfolio. Use it as an entry into your portfolio and show why viewers should stay and explore the main sections of your portfolio, which are your process and project samples.

Component 2: Your Problem-Solving Process

The highlight of your portfolio, really, should be the descriptions of your UX writing process. This typically begins with how you recognize a problem or an opportunity for improvement, where and whom you consult with for ideas, how you identify potential solutions, and what you make as the actual solution.

In this section of your portfolio, start by describing the project you were working on: What was the impetus for the project (why they needed a UX writing intervention), what was the range of the project (dates), what were the existing conditions (current or early designs), what was the specific problem you were tackling, and what were the specific roles and responsibilities you performed in this project? If you were a part of a larger team, be sure to indicate your roles and activities appropriately.

Visuals are your friends in a portfolio. Always provide screenshots, mockups, and process images to show the working process of your problem-solving. Reviewers want to know not just what you've accomplished but how you got from the initial place to the final resolution. Any decisions you made and ideas you considered would be good indicators of your problem definition and ideation abilities. When writing about your process, include the following basic information about the project:

- Project name
- Dates of project
- Problem statement
- Role(s) you played
- Activities performed

First, clarify the goal or problem you were presented with at the beginning of the project:

- Was it a redesign or a new product?
- What was the main problem?
- What caused the problem?
- What were your initial reactions to the problem?

Then, walk your reader through the decisions you made about the target audience, method selection, and the prototyping and testing process. In these detailed case studies, reviewers are looking to understand your thought process and workflow. Here are a few more tips to keep in mind:

- Think like a researcher. How did you (and your team) come up with the problem statement for your projects? How were methods selected? Who did what?
- Beyond talking about what happened in the project, share what you learned from doing it. Did you learn a new technology? Did you discover new ways to write/design? Did you found new resources?
- Most importantly, be respectful of the projects you have worked on. Even if it was a terrible design, do not trash talk about the project. You never know the background of your portfolio reviewers or the relationship they have with the projects you feature.

Component 3: Your Project Samples and Results

The final outcomes of your projects are what portfolio reviewers most desire to see. In your case studies, include results such as usability test findings, user quotes, testimonials, and business impact, if applicable (Figure 10.4). When organizing your presentation, prioritize your most impressive results first to wow your reviewer. Put another significant result at the end of a case study to leave a good final impression.

Never lie about your project outcomes, ever. If you find yourself stuck with less-than-exciting projects, which is a common issue for beginning professionals and students, do not feel pressured to amp up your projects. Tell the stories as they are, and be genuine about what you learned from them. In fact, not all real-world projects are always thrilling. You

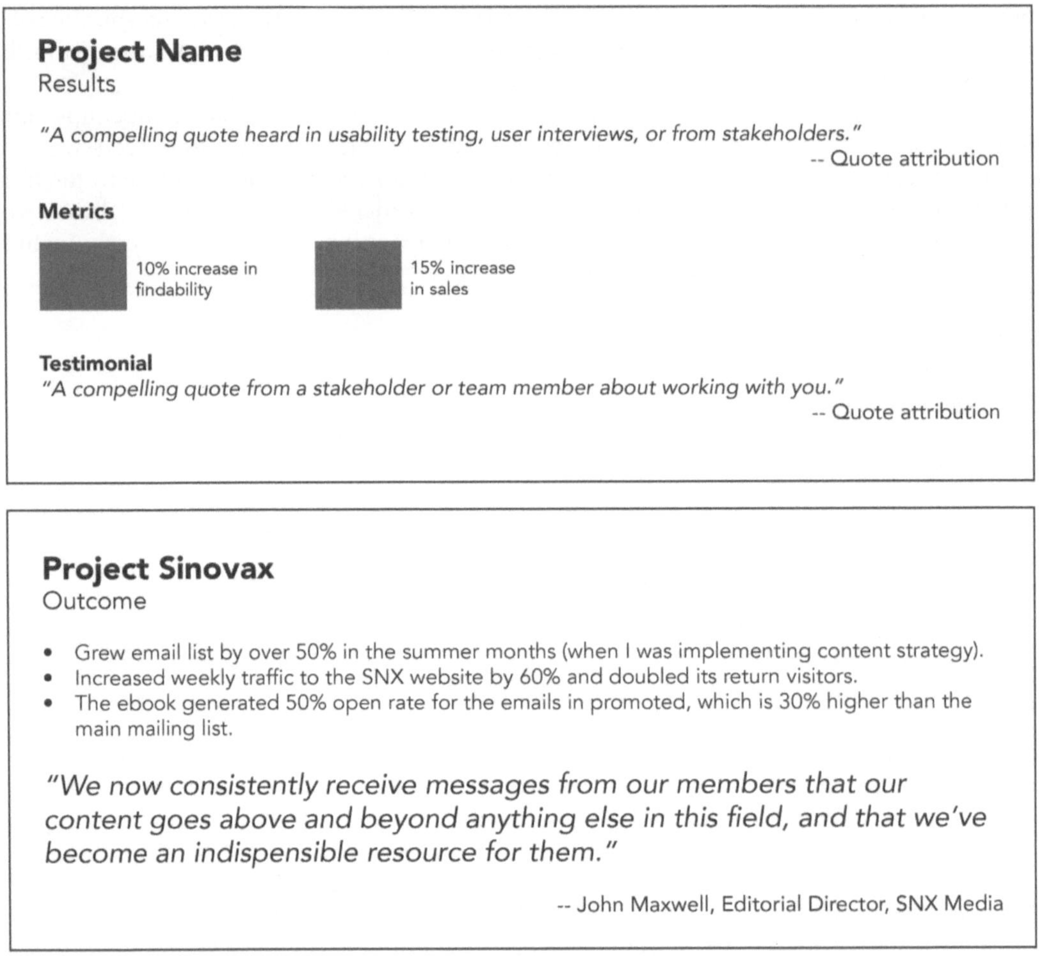

Project Name
Results

"A compelling quote heard in usability testing, user interviews, or from stakeholders."

-- Quote attribution

Metrics

10% increase in findability

15% increase in sales

Testimonial
"A compelling quote from a stakeholder or team member about working with you."

-- Quote attribution

Project Sinovax
Outcome

- Grew email list by over 50% in the summer months (when I was implementing content strategy).
- Increased weekly traffic to the SNX website by 60% and doubled its return visitors.
- The ebook generated 50% open rate for the emails in promoted, which is 30% higher than the main mailing list.

"We now consistently receive messages from our members that our content goes above and beyond anything else in this field, and that we've become an indispensible resource for them."

-- John Maxwell, Editorial Director, SNX Media

Figure 10.4 Template (top) and example (bottom) for project outcomes or results.

will encounter a lot of mundane projects on the job, so show your portfolio reviewers how good you are at managing your findings and turning them into lessons for the future.

Brand names have power. Feature projects with brand names that most people would recognize. Even if you do not work at one or have not actually been hired to do such a project, you can volunteer to pursue a redesign challenge (such as the exercises in this book) to show your ability to solve a problem within the context of a recognizable brand. If you're still short on project samples, try taking part in UX writing challenges such as those assigned on Daily UX Writing (https://dailyuxwriting.com) or playing with a random microcopy prompt generator. For instance, here is a sample prompt for creating microcopy:

Day 12, 45 characters max

A user is creating an account. When they come to the step where they are asked to enter their name, they get an error message. A fraud detection software thinks their name is fake—but it's wrong 5% of the time.

> **Write an error message that prompts them to fix the error without shaming them for having a fake-sounding name.**

Some client projects require confidentiality. If you have to deal with non-disclosure agreements, you can either blur out or redact identifying information (like names, financial or medical information, and contact information) and make your design generic (recreate the final design with different colors, logos, and images that may identify the client).

Portfolio Review

Once you have completed the skeletal content for your portfolio and feel ready to have someone look at it, prepare a portfolio presentation for a mentor, colleague, or family member. When soliciting feedback from others, be sure to inform them of your intentions or goals. That way, they may be able to concentrate on certain aspects of your portfolio and give you targeted comments. Beyond surface or technical issues, such as spelling or grammar errors, ask your peer reviewers to comment on the meaningfulness of the content and overall usability of your portfolio. You can have the peer review done synchronously or asynchronously. If done in person, you should also prepare a short oral walkthrough of your portfolio, as it is commonly done in face-to-face job interviews.

If you are new to UX writing and do not have colleagues or immediate peers who are working in the profession, you may ask an industry practitioner to conduct an informational interview with you. As you interview with them, make note of what resonates with them and what may be unclear or confusing to them. Use those notes to revise your portfolio.

Here are some comments we have gathered from UX experts who frequently review portfolios:

- "I really want to see their thinking process. The last thing I want to see is just some pretty mockups of something they have done because that doesn't tell me how this person solves problems."
- "For team-based projects, it is important to show me what you did versus what other people did on the project."
- "It's always nice to see initial sketches and how they turn into the final design."
- "If you are working on an existing product, I like to see the before-and-after comparison because it shows me visually what changes took place. Even better, include annotations to say why certain changes were made."
- "Strive for balance in the length of your content. I want to learn about the context of your project, but I don't want to spend 30 min reading the case study."

As you continue to work on new projects, make it a habit to save significant artifacts or process documents to include in future iterations of your portfolio. Keeping the working portfolio model in mind, you should create a method for keeping track of projects to make future updates easy. Your future self will thank you!

Conclusion

A portfolio is tangible evidence of your knowledge about UX writing and your skills in solving problems through design. If you haven't started building one, it's not too late to begin. Taking the first step may seem daunting, especially if you are new to the field. However, once you've set the portfolio in motion, you may find it quite simple to add content to it as you advance in your career. At any rate, always iterate and improve your portfolio by updating it with current content and gathering regular feedback from peers and expert reviewers.

Chapter Checklist

This chapter has presented the rationale and tips to guide you in creating and maintaining a UX writing portfolio. Pay particular attention to the key components of a portfolio to present a persuasive document of your skills and achievements.

- *Consider your goals and intentions for a portfolio.*
  - Create a working portfolio as a personal documentation tool. Add to it ongoingly and keep only one version to avoid confusion in record-keeping.
  - Select specific artifacts to create a presentational portfolio for particular audiences and situations.
  - Create a static portfolio to communicate your expertise quickly and directly.
  - Create a dynamic portfolio for a more immersive experience with your content.
  - Choose a social showcasing platform to host your portfolio for better discoverability.
  - Choose to host your portfolio on a personal website for better control and customization.
- *Design your portfolio to meet professional expectations.*
  - Introduce yourself authentically and provide contact information.
  - Show your problem-solving process by providing details about your projects and activities.
  - Feature the final outcomes and impacts of your projects.
- *Gather reviews for your portfolio.*
  - Invite peers and colleagues to give feedback on certain aspects of your portfolio.
  - Conduct informational interviews with industry professionals to gather expert reviews.

Discussion Questions

1 Why should UX writing professionals keep a portfolio?
2 How can you be intentional about your portfolio design?
3 Who might you count on to provide external feedback on your portfolio?

Learning Activity

If you already have a portfolio, use the recommendations in this chapter to review your existing portfolio design. Write down a list of things your portfolio is already doing, and a separate list of ideas for improving it.

If you don't have an existing professional portfolio, begin one with the following steps:

1 Locate three to five exemplar portfolios online. Note why you like them and how they embody the recommendations in this chapter.
2 Select a primary model from your examples and sketch out your own portfolio structure, such as Home, About Me, Work Samples, Contact, etc.

3 Decide on a CMS that allows you to start building your portfolio.
4 Compose and upload relevant content to your site while keeping them "unpublished" until you're ready to make them live online.
5 Share the URL of your portfolio with a trusted friend or colleague. Give them two to three things to focus on for critique and feedback. Don't ask them to comment on the whole site. Instead, pay special attention to menu labels, selected work samples, or the narrative of your personal bio.
6 Revise your portfolio design based on the feedback you collected.
7 Update your portfolio content regularly.

References

Keshtcher, Y. (2019). Best UX writing portfolios (2021 update). Medium. https://yuvalkeshtcher. medium.com/best-ux-writing-portfolios-2019-update-7e6a066631af
Williams, G. (2009). *Build your training portfolio*. The American Society for Training & Development.

11 Using Generative AI and Automating Your Content

Chapter Overview

With the introduction of generative AI, there will be changes in how content is created and distributed. UX writers will need to know how to interact with and use generative artificial intelligence (AI) through text prompts and "human-in-the-loop interventions" and develop "no-code automation" of their content. This chapter will first introduce some key generative AI programs and describe how they can work as a companion for UX writers. We then discuss how developments in the use of no-code automation can help automate content development. As UX writers will be asked to not only write but design and publish content, the use of such generative AI programs can make their job easier but also create ethical challenges. This chapter presents how generative AI functions help create hybrid content (content that is written by AI together with a UX writer) and presents an overview of how AI can help UX writers develop, iterate, and manage new content and automate repetitive tasks using RPA (robotic process automation).

Learning Objectives

- Develop an understanding of AI in general and its role in UX writing.
- Understand the duties of UX writers in working together with AI to produce hybrid content in user-friendly and ethical ways.
- Develop a practical sense of the HEAT model for UX writers to humanize AI-written content.
- Learn about different generative AI programs and what they can do.
- Understand how generative AI can generate different components relevant to UX writing.
- Understand how generative AI and no-code automation create automatic, dynamic content management solutions for UX writers.
- Practice the ethics of using AI in your work.

What Is Artificial Intelligence, Again?

During the writing of this book, developments were taking place in AI that we wanted to address and situate in relation to the activities of UX writers. The development of AI is part and parcel of the development of computers and systematic ways of thinking and has

DOI: 10.4324/9781003274414-14

been for years. However, recent developments in AI also impact how UX writers create and distribute content. AI programs can now generate text, images, videos, and music as content. According to researchers Ann Hill Duin and Isabel Pedersen (2021), AI can now function as an *augmentation technology*. AI extends, and thus augments, human capabilities by automatically generating text, images, videos, or music (that a human user takes hours, if not days, to create) in just a few seconds. According to Duin and Pedersen, we are now looking at activities where AI automatically generates work as a *companion* to UX writers in generating content using a chatbot-like interface.

For example, generative AI writing bots such as OpenAI.com's Chat Generative Pre-Trained Transformer (ChatGPT), Jasper.ai, Sudowrite.com, or Rytr.me can quickly generate any type of long or short text-based answer on a simple text prompt from a user with a few keywords. However, these tools do not always understand the specific audience and will be creative with facts, so they still need a UX writer to bring the human perspective to create output that works for a specific audience.

Generative AI is certainly not without its controversies. As we mentioned earlier in the book, the Gartner hype cycle holds that the early embrace of new technologies often leads to speculation and excitement for innovative uses of that technology before a more accurate, realistic view is developed as reality sets in for users as to what the technology can actually do well. Generative AI is no different. Many issues arise from the fact that AI bots need to be "trained" to create the content they generate. They learn from the databases on which they're trained, and if those databases contain inaccurate, biased, libelous, or even scandalous material, then the bots will reproduce that content unless corrected and fact-checked by humans.

AI bots can also and do generate content that may be the intellectual property of others. For example, visual designers and artists have sued AI companies for using their artistic work without their permission in the training set for text-to-image generators such as Midjourney and DALL-E (Vincent, 2023). In 2023, a scandal erupted when technology website CNET was exposed for using AI bots to write its content. It was found that CNET AI bots generated plagiarized posts from other sites with incorrect information, forcing the website to put warnings on all of them and causing a loss in credibility with their readers (Christian, 2023). Such uncritical use of AI writing bots can have serious repercussions for the credibility of an organization since the content it generates is not technically "new" and is not fact-checked or accurate.

Others have criticized how generative AI text programs (such as ChatGPT) may unwittingly perpetuate harmful ideas by uncritically using biased training data to generate racist, sexist, or discriminatory content that is harmful to people without questioning (Noble, 2018; Akter et al., 2021; Hocutt et al., 2023). And, generative AI programs such as ChatGPT have been criticized for being trained on only a small amount of gigabytes of data, creating only a "blurry" image of the complexity of the web that lets AI do mere "paraphrasing" of written work and ideas that are already on the Internet (Chiang, 2023). Limited training of AI writing bots can thus have large-scale repercussions since AI gives the impression of knowing what it is writing about.

Another problem with generative AI is that it sometimes comes up with incorrect answers (so-called "hallucinating" answers). AI may generate incorrect output without any understanding of whether the answers or writing it has created are factually correct. For example, in February 2023, Google lost $100 billion in its valuation when, in a public demo, their AI chatbot Bard incorrectly identified the James Webb Space Telescope as the first telescope to take images of planets outside our solar system (Coulter & Bensinger, 2023). Unfortunately, the correct answer was that the European Southern Observatory's Very Large Telescope had already done this and was the first to do so in 2004. AI content should thus never be taken at face value and always be checked for accuracy. So an important issue for UX writers when looking at AI-developed content is the need to always check the content for accuracy.

Commercial use of generative AI may also create differences in the quality of output as tiered plans and access to AI may be introduced based on how much people are willing to pay on a monthly basis for AI services. A user who pays for generative AI may receive more in-depth answers, whereas a free service user may get less in-depth information and quality of answers. Hence, it is important that UX writers develop an ethical framework for working with AI that involves cross-checking the accuracy of content and *humanizing* AI-written content by adding their human touch. Nevertheless, despite above shortcomings of generative AI, there are interesting developments that you will need to consider as a UX writer, which we describe below.

Developments in generative AI writing have led to a revolution in the writing of prompts to help generate specific texts. Writing prompts for AI can be simple, such as "write me a cover letter," or they can be specific: "write a cover letter for a UX writing position with five years of experience in human-computer interaction." What's more, using a chatbot-like, conversational interface, generative AI writing bots can take input from users through dialogue to refine what has been written and so iterate on new output. Hence, a user can write a follow-up prompt to the chatbot and state, "now shorten the cover letter to 500 words," and the AI's output for the cover letter for the UX writing position is now shortened to 500 words. Such curation and refining of text can significantly help UX writers think through their work and generate new ideas very quickly using AI as a writing companion.

AI writing bots are very versatile, too. They can write specialist content such as "recommend titles for blog posts on specific topics," write said blog posts, and provide search engine optimization (SEO) meta tags for that blog post so search engines list the content it has written prominently. They can even write whole social media campaigns. AI writing bots can also deal with tasks of increased complexity, such as writing specific computer code to run programs and incorporating specific design thinking or product management techniques into prompts to come up with original ideas for UX writers to use. For example, you can ask an AI writing bot to "create a journey map for an innovative fitness app and use tabular format," and it will spit out a journey map in a table with suggestions for your fitness app.

AI bots also help UX writers do quick user research and sentiment analysis. For example, AI bots can extract frequent keywords from text, which is useful if you give them user reviews to analyze for the top ten most commonly occurring keywords. For a UX writer quickly trying to get an idea of how users are evaluating your content, this can be very insightful to get a feel of what your users are saying about your UX content, product, or service. Likewise, AI writing bots can analyze negative or positive sentiments from user reviews. For instance, by using the prompt to the AI to "use sentiment analysis and indicate if the below review is positive or negative"—and giving the AI a user review such as, "this is a great product!"—it will answer "positive sentiment" because the user is feeling positive about the product. Again, for UX writers, AI bots can be very handy in doing preliminary user research in a very quick manner. However, you will want to do some checks and see if what they are categorizing is truly correct, as AI may make mistakes sometimes.

AI bots can also help clean up and analyze data, which is useful to sort user data. AI bots can be used to do basic data analytics operations, such as data cleansing (like changing date formats from European 23/12/2024 to US formats 12/23/2024, or removing elements such as emojis from large reams of text). AI bots can also use conditional logic to quickly categorize elements in data (such as "if a number is below 5, then write 5, if it is higher, then write 5–10"), which can help analyze and sort data quickly into categories.

At last, AI bots can also help come up with new ideas in an iterative manner. AI writing bots can also write complete scripts for YouTube videos explaining complex topics, which can be really handy if you are trying to learn new technologies. And, as mentioned, when the user tells the bot to change something, the AI writing bot can take in that information and learn from it while generating new text output iteratively. As mentioned, AI bots respond to UX writer prompts and can refine their answers in an iterative manner through a chat-like dialogue mechanism, so they work in tandem with UX writers.

The Limitations of AI and the Importance of Human-in-the-Loop Approaches for UX Writing

Not all of the text generated by AI is correct or desired. As we noted before, AI is prone to hallucinate, in that it sometimes makes stuff up. For example, when we asked it to give us the first, third, and fifth words of the following sentence: "Your notebook looks great," ChatGPT gave this incorrect answer: "Your, looks, great." If you counted the number of words in this sentence, you'd realize it only has four words, so the last suggestion ("great") is incorrect. Likewise, ChatGPT has difficulty with logic like this: "If Rose has four sisters and Bob is her brother, how many sisters does Bob have?"—it will incorrectly answer that "Bob has four sisters" and forget that Rose is also a sister. For this reason, generative AI will require that a UX writer check the work of the AI for accuracy and whether it has gotten certain facts right. Future versions of AI may correct these issues, as AI is learning from a growing dataset. But for now, UX writers must validate the accuracy of AI content.

AI-written content may also lack a human dimension. Companies such as Google have cracked down on AI-only written content because it can be quickly generated and thus create problems for the accuracy of information. This has large-scale consequences since bad actors can quickly generate convincing texts but use them to spread misinformation. For this reason, Google is penalizing websites that feature AI-only written text. According to technologist Neil Patel:

> Google doesn't like AI-written content, but it does like AI writers. What Google wants you to do is take the AI content and use it as a starting place and modify it. Take that AI-written content and make it better—provide more value.

For this reason, Patel recommends writers add their expertise, authority, and trust as well as humanize the AI-written content by modifying it so it works for a human user (Visual Stories, 2023).

In our case, we simply call this mechanism as UX writers bringing the **HEAT** to perhaps cold, anemic AI-written content (and bringing their Human experience, Ethics, Authenticity, and Trust to it). Similar to a self-driving car, it is still important to have a human operator check or be able to intervene to avoid any accidents. UX writers need to check their work and be able to justify the choices they have made to their employer when addressing their audience.

For example, using the HEAT model, a UX writer can create better hybrid content if they apply the following elements to AI-written content:

- **Human Experience**: Check if the information is factually correct; also check if the content generated is biased toward specific demographics of users or diminishes specific groups of users, and correct the human user experience.

- **Ethics:** Check if the user experience in the content is promoted in a way that is ethical and user-friendly, provides the user with a positive experience, and informs them how their data are being used to protect their privacy and promote positive experiences.
- **Authenticity:** Personalize the content, add references, citations, and links so the information can be checked independently by users and yourself to be deemed authentic rather than created by an AI bot.
- **Trust:** Make sure to build trust and add the human dimension (think of how the information may impact your user, whether negatively or positively), and explain what can be done to achieve better outcomes. Provide ways to reach out to you personally via social media and email to build trust with your audience.

This chapter advocates that AI and UX writing go hand in hand, but it does require the UX writer to incorporate a "human-in-the-loop" approach to using AI responsibly in their work. That is, UX writers need to modify AI-written content to better address their audience and be able to justify their choices from a human perspective. As any data used for AI training may be biased or inaccurate, AI researchers such as Ge Wang (2019) at the Stanford University Human-Centered Artificial Intelligence lab have noted the importance of human-in-the-loop approaches to fostering better outcomes with AI (see Figure 11.1).

According to Wang (2019), the benefits of human-in-the-loop approaches to AI are that they can create better user experiences and more human-friendly, ethical, and energy-friendly AI.

- **Transparency:** Because a human is involved, the AI system is meant to work alongside humans, making AI more transparent in how decisions are made. This is important since AI needs to be accountable to humans, not run autonomously on its own and "black box" its decision-making. A bonus is that transparency makes for better, explainable AI where end-users understand what the AI is doing.

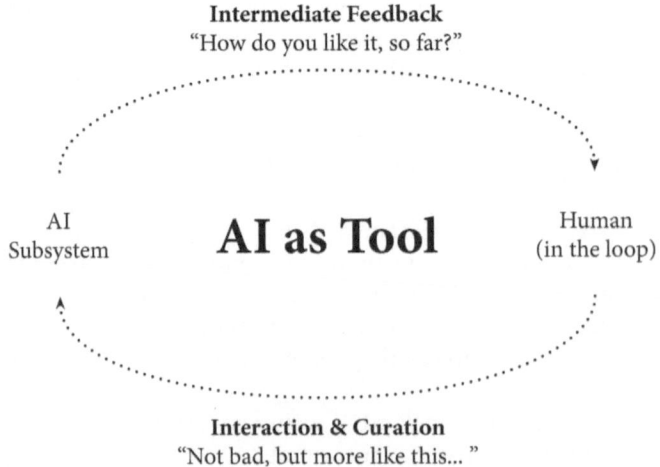

Figure 11.1 A human-in-the-loop approach to interacting with AI requires humans to intervene.
Source: Adapted from Wang (2019).

- **Incorporation of human judgment:** Human values are incorporated into an AI system, which makes it work better since AI has to incorporate the nuances of underlying human emotions and motivations by having a human check the work of AI.
- **Simplifying algorithms:** Human involvement can also create better-functioning AI since they can determine if simpler AI programs work better in certain cases. Instead of using complex AI that tries to do everything and uses unnecessary computing power (wasteful energy use) or isn't really necessary for the case at hand, humans can decide simpler AI programs are sufficient.
- **Better overall functioning AI systems:** Human input can make for a better overall AI system since it learns from humans where it should help and where it should hold off and allow the human user to do the work (Wang, 2019).

In sum, AI alone won't do the job; UX writers still need to be able to justify why they have made certain design choices and make ethical choices that will help their audience do things in a user-friendly manner, using the HEAT model and human-in-the-loop perspectives.

Lucky for UX writers, AI has progressed so much that it can now incorporate a chat dialogue system where writers can modify the text generated by AI in an ongoing, dynamic manner. In advocating for content to be dynamically written by AI and modified by a UX writer to fit the needs of their audience, we believe AI can indeed lead to better solutions for users. And by adding HEAT to AI-generated content, UX writers play an important role in curating and modifying automated content that would otherwise be anonymous and unfit for users. The role of UX writers is thus intrinsic in making sure hybrid content works for the audience, is factually correct as well as unbiased, and for ensuring that the content works for that audience.

Since AI can mimic humans in producing certain content but does not actually know how the content is received by an audience, it does not have a clue about the success (or failure) of the given content. In fact, ChatGPT admits that "while we (OpenAI.com) have safeguards in place, the system may occasionally generate incorrect or misleading information and produce offensive or biased content." However, as AI writing bots are trained using big data and are constantly learning from human users, they will continue to improve. Nevertheless, UX writers will still need to be on the lookout for biased, incorrect, or inappropriate content produced by AI.

Using Generative AI: A Demo

While AI may have issues, AI writing bots can help UX writers with many tasks. Many generative AI systems essentially work like chatbots with which UX writers can hold a dialogue and think through different scenarios and design decisions. A UX writer can converse with AI, modify the prompt, and thus iteratively refine what the AI chatbot writes for them. Again, this happens at breakneck speeds. The following scenario (Figures 11.2–11.6) took the generative AI only a matter of minutes to complete multiple queries. In that time, AI-generated 10 blog post ideas for titles, wrote the first blog post, rewrote the blog post in the folksy style of a famous writer (Mark Twain), and provided SEO keywords and a potential URL that will hopefully play nice with search engines. The possibilities of using AI bots to do UX writing tasks, as the example shows, can speed up your work.

To demonstrate, a UX writer might ask ChatGPT, "give me ten blog post titles on UX and intercultural communication," and the AI will spit out exactly 10 titles in a few seconds (Figure 11.2).

Then, a UX writer can follow up by picking one of the given titles and asking, "Write the actual blog post for '5 Tips for Improving Intercultural Communication in UX Design.'" With that, ChatGPT will generate the actual blog post in a similar short time (Figure 11.3).

If the post sounds too serious, a UX writer can ask ChatGPT to write in a particular famous person's style (for instance, "Now rewrite the post in folksy Mark Twain style." See Figure 11.4).

Obviously, UX writers should add their own perspectives (i.e., applying HEAT) since the post is somewhat barebones. They may want to add sources where the information came from, such as mentioning the work of researchers such as Geert Hofstede and Erin Meyer—well-known scholars in intercultural communication—along with UX scholars or researchers. However, if the UX writer likes the blog post, they can ask the generative AI to suggest how to provide metadata that helps with SEO, so the blog post can be picked up by search engines and promoted (see Figure 11.5).

And that is not it. A UX writer can even ask ChatGPT to recommend a specific optimized URL that can be easily picked up by browsers, such as "www.yourwebsite.com/5-tips-intercultural-communication-ux-design-mark-twain-style" (Figure 11.6).

If the UX writer doesn't find the first response good enough, they can ask the generative AI to "regenerate response" and so write a different version of the text along the same idea. Rinse and repeat.

With AI chatbots and their ability to automatically generate specific texts based on shifting user prompts, one can see that UX writers can do a lot with generative AI to do a large amount of work in a very short time. And because generative AI writing bots save

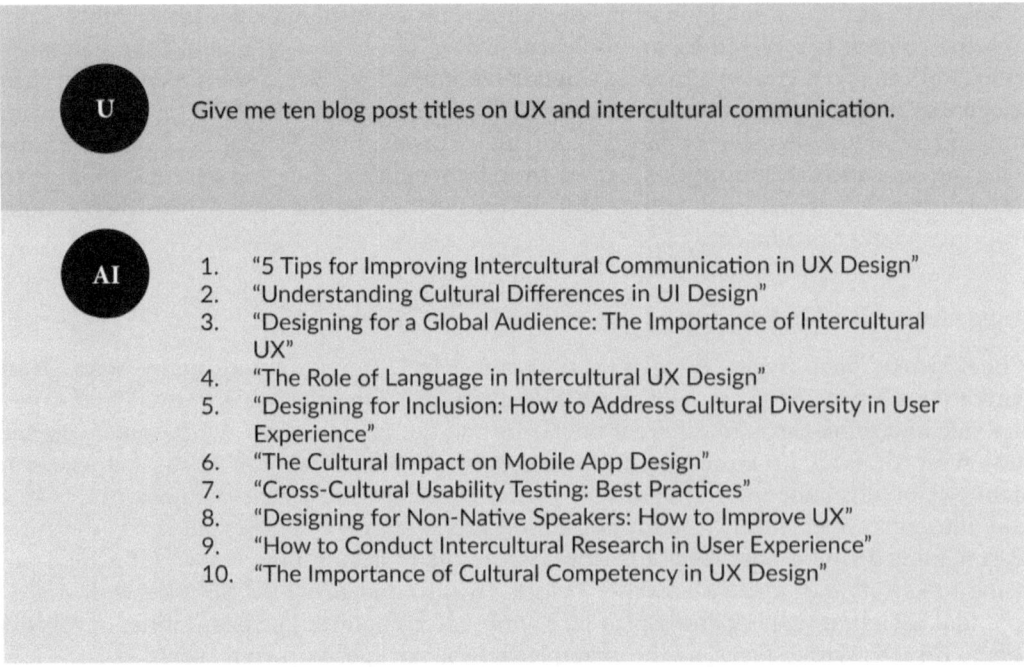

Figure 11.2 Generative AI writing bot ChatGPT generated blog post titles for the prompt, "Give me ten blog post titles on UX and intercultural communication."

Write the actual blog post for "5 Tips for Improving Intercultural Communication in UX Design"

As our world becomes increasingly connected and globalized, it's more important than ever for UX designers to understand the role of intercultural communication in their work. Whether you're designing a website or mobile app that will be used by people from all over the world, or working with a team that spans multiple cultures, understanding how to communicate effectively across cultures is essential. In this blog post, we'll share five tips for improving intercultural communication in UX design.

1. Understand your own cultural perspective. The first step in improving intercultural communication is to understand your own cultural perspective. This means recognizing the values, beliefs, and assumptions that shape the way you communicate and interact with others. By being aware of your own cultural background, you'll be better equipped to understand and communicate with people from other cultures.

2. Learn about the cultures you're designing for. To effectively communicate with people from other cultures, it's essential to understand the values, beliefs, and communication styles that shape their perspectives. This means taking the time to research the cultures you're designing for, and learning as much as you can about their customs, traditions, and ways of communicating.

3. Be mindful of language barriers. Language barriers can be a major obstacle to effective intercultural communication. To overcome this, it's important to be mindful of the languages spoken by your target audience, and to consider ways to make your design accessible to non-native speakers. This might include providing translation options, or using simple, easy-to-understand language in your design.

4. Be open to feedback and adapt your design accordingly. In any design process, it's essential to be open to feedback and willing to make changes as needed. This is especially important in intercultural communication, where cultural differences can lead to misunderstandings or misinterpretations. By being open to feedback and willing to adapt your design, you'll be better able to create a user experience that resonates with people from all over the world.

5. Embrace cultural diversity in your team. Finally, one of the best ways to improve intercultural communication in UX design is to embrace cultural diversity in your team. This means actively seeking out team members from different backgrounds and cultures, and creating an inclusive environment that values different perspectives. With a diverse team, you'll be better able to understand and communicate with people from all over the world.

In conclusion, intercultural communication is an essential aspect of UX design, and it's important for designers to understand the role of culture in shaping the way people communicate and interact with technology. By following these tips and being open to feedback, you'll be better equipped to create a user experience that resonates with people from all over the world.

Figure 11.3 ChatGPT writes the actual blog post using the title it originally generated.

U Now rewrite the post in folksy Mark Twain style.

AI Well folks, it's a big ol' world out there and it's gettin' smaller every day. With all the newfangled technology and whatnot, we're all connected like never before. And you know what that means for us UX designers? It means we gotta start thinkin' about how we communicate with folks from all over the world.

Now, I ain't no expert on the matter, but I've picked up a thing or two in my travels. So, I thought I'd share with y'all 5 tips on how to improve yer intercultural communication in UX design.

1. Get to know yerself. The first step in understandin' other folks is to understand yerself. Take a good look at the values, beliefs, and assumptions that shape the way you communicate and interact with others. By knowin' yerself, you'll be better equipped to understand and communicate with folks from other cultures.
2. Learn 'bout the folks you're designin' for. If you wanna effectively communicate with folks from other cultures, you gotta take the time to learn 'bout 'em. Research their customs, traditions, and ways of communicating. The more you know, the better you'll be able to connect with 'em.
3. Watch out for them language barriers. Language barriers can be a real pain when it comes to intercultural communication. To avoid confusion, pay attention to the languages spoken by yer target audience and consider ways to make yer design accessible to non-native speakers.
4. Be open to feedback and adapt yer design accordingly. Now, listen here folks, in any design process it's important to be open to feedback and willing to make changes as needed. This is especially true when it comes to intercultural communication. Be open to feedback and willing to adapt yer design and you'll be able to create a user experience that folks from all over the world can relate to.
5. Embrace cultural diversity in yer team. Finally, one of the best ways to improve intercultural communication in UX design is to embrace cultural diversity in yer team. Seek out team members from different backgrounds and cultures and create an inclusive environment that values different perspectives. With a diverse team, you'll be better able to understand and communicate with folks from all over the world.

In conclusion, intercultural communication is a mighty important aspect of UX design and it's crucial for us designers to understand the role of culture in shaping the way folks communicate and interact with technology. By following these tips and being open to feedback, you'll be better equipped to create a user experience that folks from all over the world can enjoy.

Figure 11.4 ChatGPT rewriting the previous blog post in the folksy style of American author Mark Twain.

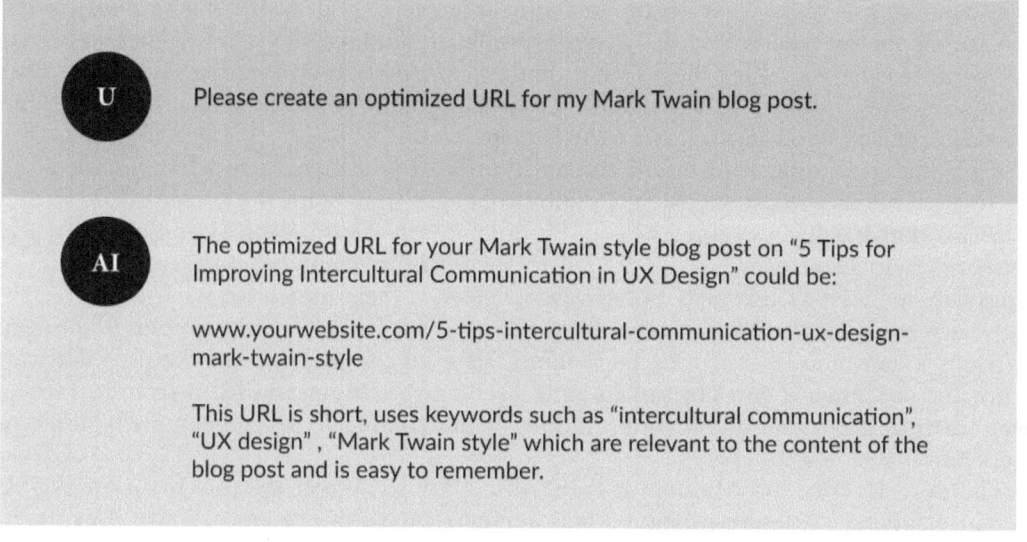

U Please generate SEO keywords for this post.

AI
1. Intercultural communication
2. UX design
3. Cultural diversity
4. Globalized communication
5. User experience
6. Cultural perspective
7. Language barriers
8. Inclusive design
9. Cross-cultural communication
10. Multicultural team
11. Intercultural UX
12. User-centered design
13. International communication
14. Cultural sensitivity
15. Design for global audience
16. Cross-cultural usability
17. Intercultural user research
18. Global design
19. Multilingual design
20. Inclusive user interface

Figure 11.5 ChatGPT recommends SEO keywords for metadata that help promote the blog post via search engines.

U Please create an optimized URL for my Mark Twain blog post.

AI The optimized URL for your Mark Twain style blog post on "5 Tips for Improving Intercultural Communication in UX Design" could be:

www.yourwebsite.com/5-tips-intercultural-communication-ux-design-mark-twain-style

This URL is short, uses keywords such as "intercultural communication", "UX design" , "Mark Twain style" which are relevant to the content of the blog post and is easy to remember.

Figure 11.6 ChatGPT's recommendation for a URL that can work for said Mark Twain-style blog post.

such interactions in conversation threads (or interaction histories), UX writers can revisit the work they have done with ChatGPT at any time.

How Generative AI Actually Function

You may be asking yourself: How do generative AI bots function, do all of this work, and respond in such a lifelike fashion to the demands of UX writers? Well, think of generative AI like the act of peeling the layers of an onion. Each layer in an AI adds something that creates the illusion that you are talking to an intelligent person. For example, the development of "deep learning" and "neural networks" techniques in AI create powerful learning capabilities that help AI create engaging UX content as output that learns from text input from a user (Hocutt, Ranade & Verhulsdonck, 2023). Natural language processing (NLP) and natural language generation (NLG) techniques in AI help understand what the user is asking for and generate answers to their queries. And large-language AI models are trained to write on specific topics and find meanings between different topics.

"Deep learning" essentially identifies patterns from many, many reams of data (such as texts) and continuously develops a model. According to IBM (2020), deep learning is "a subset of machine learning" that uses neural networks to learn which inputs (say, a voice command from a user) lead to which specific successful outputs (giving a recommendation that the user likes). Deep learning is especially important because we know our users want positive experiences and technologies that remember their personal preferences and respond to voice input.

Hence, AI bots use deep learning to learn what a typical user likes and which answers work with which prompt. The neural network is simply a way of finding which questions lead to successful answers when the user signals they are satisfied.

A second component of our AI-onion metaphor is the ability of AI to converse with you as the user. AI does this through NLP. For example, when you give a voice command to your digital assistant (like Siri or Alexa) for "Find Mexican food near me," AI translates what you queried into "use my GPS location to identify Mexican restaurants near me and tell me about them." NLP uses two techniques to discern what you are saying: finding your *intent* (why you are asking or what action you want to achieve) and identifying an *entity* (what you are asking for, specifically). So, in our example of "finding Mexican food near me," the user's intent is to ask AI for the action of finding information on locating restaurants near them using the entity (or item/criterion) of Mexican food. As such, the AI processes: "**Find** {*Intent*} **Mexican Food** {*Entity*} **Near Me** {*Intent*}".

After understanding your intent and entity, the AI will then parse what you said and try to formulate a response using a dialog manager that pairs the user's query with a response that has been deemed helpful by other users. Again, deep learning and neural networks help identify which inputs led to successful outputs for different users. Since input can be messy (a user may not always be clear in their speech for what they are asking), deep learning also tries to guess what the meaning is in the unclear input. Since deep learning is continuously learning from inputs, if it gets feedback from the user that this is not the outcome, it will put this in its ruleset for the future and learn from it. Hence, deep learning has major implications in that it can dynamically learn from many different users and ingest new rules.

The third layer of the AI onion is then using NLG to answer the user in an intelligible format. In giving a recommendation that is intelligible to us (for instance, Siri or Alexa says, "the nearest Mexican restaurant is Chachi's, which is 6 minutes' drive away from your current location"), the AI is using NLG. Coupled with its large language model capabilities, AI bots can thus not only infer what the user wants but also formulate an intelligible speech-like response as output. And, as we noted above, it can do so in seconds. And generative AI writing bots can respond to user text prompts and write on any topic in any format, style, or manner you like.

The above response from an AI-driven digital assistant chatbot is simple and straightforward. But what happens if your question about AI is much more vague or involves creating something new? Developments in generative AI have created a revolution in how users can now interface with the world and create new content. As generative AI can generate texts, images, and interfaces in seconds, UX writers and content designers are looking at how this can create new automated workflows where AI does the heavy lifting. The role of generative AI thus augments the work of UX writers by vastly speeding up the creation of content.

Text-to-image AI services like Stable Diffusion, Midjourney, and DALL-E can create new images using simple text prompts from users. Again, here is where NLP is featured prominently because the AI system has to translate what you say (in written text) to a desired image. Text-to-image AI also functions by users feeding them prompts of text, which they translate and generate visuals using billions of images as a training set. Rather than a human user slaving away in a design program like Adobe Illustrator to create one graphic and work on it for hours on end, generative AI can create the required graphic in seconds, thus creating opportunities for people to do jobs that were previously specialized to artists.

The use of such generative AI is not without its controversies, of course, as artists have argued that the technology is based on large training sets that use specific artists' work and thus violates their copyright by using their work without their permission (Vincent, 2023). Likewise, companies may object to using AI-generated content simply because they cannot copyright it. Furthermore, the use of images to create just about anything visually can have repercussions if users create photorealistic images that are used to create misinformation or propaganda. Using such images can thus lead to further removal of the ability to distinguish between what is artificial and what is an actual photo of an event. The importance of UX writers making ethical choices in these situations cannot be overstated.

Automating Your UX Writing Content: Different Tools

As we noted so far, generative AI uses a variety of AI elements to create content and can create text, images, and even music. Next to generating new content, AI can also help your workflow as a UX writer. UX writers work in teams and use content management systems (CMS) to develop a content strategy. UX writers will need to develop content calendars to work with and schedule when omnichannel content needs to be pushed out to social media, websites, newsletters, and so forth. Here is where automating your UX writing also becomes important.

Sophisticated component CMSs such as Darwin Information Type Architecture (refer to Chapter 8) can combine text and images, push out content, and create websites and PDFs using one markup language (i.e., XML). Nowadays, many UX writers are using Figma, a collaborative web application for interface design that lets them work on interfaces in one shared interface and share their work directly with colleagues who may be focused on different components of the user experience. Conversational platforms like Slack or Microsoft Teams help them communicate across an organization and share files in threads of conversation with colleagues.

Now, what if a UX writer wants to automate some of the tasks they are doing using AI that are repetitive? As noted before in this book, UX writers use content calendars and CMSs to help streamline their work by specifying when something is due and when it needs to be pushed out by the CMS. By setting dates for times when new posts are made to social media accounts as part of a larger social media campaign, UX writers know when they have to post something. However, when posting the same information on several different platforms, such tasks can get repetitive. A UX writer may need to write said social media post, then go to various social media accounts (Facebook, Twitter, and LinkedIn) and copy and paste their content, such as text and images. What if AI could help them do this at the click of button using a centralized service such as Google Sheets?

The development of no-code automation systems such as Celonis' Make.com, Microsoft Power Automate, or even UiPath can follow and create action flows that help make repetitive work by UX writers easy. For example, Make.com uses a drag-and-drop interface that can help a UX writer automate a task. Say a UX writer wants to create a social media post in Google Sheets and then push the post out to several social media accounts with one click. They can then create an automatic workflow by linking these repetitive steps into an *action flow* using Make.com. Here is where RPA can help UX writers do this task "auto-magically."

Here, no-code automation can be very powerful in helping UX writers create faster workflows by integrating AI into their work processes (Figure 11.7). No-code automation is based on simple drag-and-drop tools that allow you to automate your workflow and eliminate repetitive tasks with little to no coding involved. The idea of such automation tools is based on the idea of actions that occur and triggers that set those actions in place. For example, you could define an action in a place where if someone posts on your social media account, a record of that is saved in a database. The trigger is essentially that when the change in your database is received, it sends out an automatic message in, say, Slack,

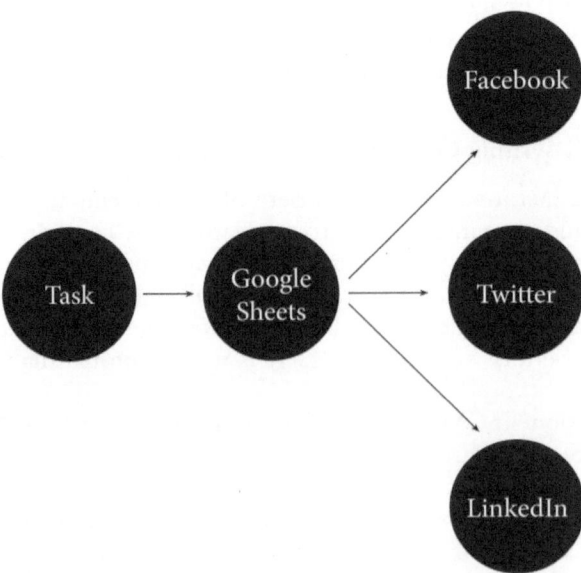

Figure 11.7 An automated workflow in Make.com allows one to create a post in Google Sheets and push it out to several social media accounts with a single click using process automation.

like, "Hey, you have a new post in your social media account." If you think about it, such scenarios can make the work of UX writers easier because they help automate workflows based on relevant scenarios. It also helps that no-code automation programs feature easy integration with apps and services such as Google Drive, Gmail, Slack, Twitter, LinkedIn, Facebook, and so forth. Connecting with those services thus creates a much quicker way of doing things using a centralized automation tool.

Here are some other tools that UX writers are using to create content:

- **Figma:** A collaborative interface design program that lets UX writers collaborate by sharing the same design. This tool is ideal for collaborative dynamics because different members of the same UX team can focus on components such as the user interface design, microcopy, or error messages in the same environment.
- **Slack:** A conversational platform that lets people organize in channels, share files instantly, and keep a record of this. It is a workplace favorite for many people since you can have messaging, conversation logs, and shared files in specific channels.
- **Airtable:** A spreadsheet-database hybrid that captures not only text but also images. Airtable is ideal for keeping components together in a logical structure, such as having a blog post and the image you plan to use for that blog post in the next column. Using such assets together creates easier content management and scheduling.
- **Placid:** A software that can flexibly create dynamic templates for text and images. Using Placid, a UX writer can create a template and dynamically generate different titles and images that look uniform.

If you think about it, using AI and automating your content can create a scenario as follows: You can create one centralized sheet (using Airtable) to store images and text and use generative AI to automatically write a blog post, title, and create an image based on the blog post. Then, you can use a no-code automation tool such as Make.com to pull that information into a dynamic template using Placid, automatically generate a title and image for the blog post, and then post that on different social media accounts with the click of a button. The possibilities are almost limitless.

Conclusion

In this chapter, you have learned about the developments in generative AI. Generative AI can produce text, images, videos, or even music in a very short amount of time using prompts given by a human operator. You have learned how generative AI programs like ChatGPT, Sudowrite, and Rytr.me can respond to text prompts and write responses to user requests in seconds. We have demonstrated how generative AI uses a chat-like interface and can take in user feedback to develop more fine-tuned content by modifying previous responses to a user. Using the HEAT guidelines, you may recognize that human-in-the-loop approaches are important to create hybrid content written by AI and checked by a UX writer to include human experience, ethics, accuracy, and build trust with the end users.

This chapter also covered the importance of how generative AI, when combined with simple no-code automation tools, can create an even faster workflow by introducing RPA to common repetitive tasks. By dragging and dropping different apps into a workflow and using a tool such as Make.com or Microsoft Power Automate, UX writers can automate certain repetitive steps (such as having to manually copy and paste a social media post into different social media accounts). The ecosystem of apps these tools can make use of when automating certain repetitive tasks provides a good way of integrating AI into a UX writer's everyday activities. Hence, with automated content using AI, including tools where AI carries the load for certain manual tasks, UX writers can do a lot in a small amount of time. This creates opportunities and benefits for working in a faster and more efficient manner. Yet it still needs an understanding of different scenarios in terms of how UX writers can use AI effectively, which is what we will discuss in the next chapter.

Chapter Checklist

This chapter has covered the implications of generative AI for UX writing activities, with specific recommendations for using AI bots to support content creation and design.

- *Get generative AI to automate parts of UX writing.*
  - Use text prompts to instruct AI bots about content needs.
  - Enhance automated content with unbiased and accurate information.
- *Use HEAT to humanize AI-generated content.*
  - Apply human-in-the-loop approaches to intervene and justify content choices.
  - Ensure content is trustworthy, accurate, and believable.
- *Apply RPA to improve the UX writing workflow.*
  - Develop no-code automation to speed up UX writing.
  - Learn different applications that allow apps/services integration to maximize hybrid content quality.

Discussion Questions

1 What is your understanding of the history of AI?
2 How are current AI systems similar to or different from voice-interacting assistants like Siri and Alexa?
3 How might AI support UX writing activities? Give some examples.
4 In what ways are generative AI applications changing UX writing?
5 What do you envision as the future of UX writing in the presence of AI?

Learning Activity

Compare two to three generative AIs that are available online: ChatGPT, Sudowrite, and Rytr.me. How are they similar or different? Give them the same prompt and compare their output. Again, what's similar and what's different? How would you evaluate the quality of the output?

References

Akter, S., McCarthy, G., Sajib, S., Michael, K., Dwivedi, Y. K., D'Ambra, J., & Shen, K. N. (2021). Algorithmic bias in data-driven innovation in the age of AI. *International Journal of Information Management, 60*, 1–13. https://doi.org/10.1016/j.ijinfomgt.2021.102387

Chiang, T. (2023). ChatGPT is a blurry JPG of the Web. *The New Yorker.* https://www.newyorker.com/tech/annals-of-technology/chatgpt-is-a-blurry-jpeg-of-the-web

Christian, J. (2023). CNET's AI journalist appears to have committed extensive plagiarism. https://futurism.com/cnet-ai-plagiarism

Coulter, M, & Bensinger, G. (2023). Alphabet shares dive after Google AI chatbot Bard flubs answer in ad. https://www.reuters.com/technology/google-ai-chatbot-bard-offers-inaccurate-information-company-ad-2023-02-08/

Duin, A. H., & Pedersen, I. (2021). *Writing futures: Collaborative, algorithmic, autonomous*. Springer.

Hocutt, D., Ranade, N, & Verhulsdonck, G. (2023). Localizing content: The roles of technical & professional communicators and machine learning in personalized chatbot responses. *Technical Communication*. https://doi.org/10.55177/tc148396

International Business Machines. (2020). AI vs. machine learning vs. deep learning vs. neural networks: What's the difference? https://www.ibm.com/cloud/blog/Ai-vs-machine-learning-vs-deep-learning-vs-neural-networks

Noble, S. U. (2018). *Algorithms of oppression: How search engines reinforce racism*. New York University Press.

Vincent, J. (2023). AI art tools Stable Diffusion and Midjourney targeted with copyright lawsuit. https://www.theverge.com/2023/1/16/23557098/generative-ai-art-copyright-legal-lawsuit-stable-diffusion-midjourney-deviantart

Visual Stories. (2023). Unlock the secrets to humanize your AI-generated content and safeguard your website. https://visualstories.com/info/humanize

Wang, G. (2019). Humans in the loop: The design of interactive AI systems. https://hai.stanford.edu/news/humans-loop-design-interactive-ai-systems

12 AI Recipes for UX Writing

Chapter Overview

This chapter shows how UX content can be generated by artificial intelligence (AI) through the use of specific scenarios. As AI responds to specific prompts—text commands by UX writers—it is important to learn the different aspects of writing such prompts. AI can suggest, augment, and modify its output quickly based on specific command prompts and do so in an iterative, conversational manner when prompted by a user. Similar to cooking recipes, writing effective prompt commands is important to the work of UX writers so they can develop better scenarios and augment and modify their own work through the use of AI. The ability of UX writers to understand how to create such recipes for the success of their content will be discussed here. These recipes include generating successful UX writing tasks such as developing personas, generating journey maps, and creating concepts for UX, as well as incorporating specific commands for AI to take specific actions and create output content for UX writing purposes.

Learning Objectives

- Develop an understanding of effective prompt writing for developing UX content through the 6W+1H question method.
- Differentiate between artificial intelligence (AI) actions and desired output.
- Implement the output-first method to think about the overall goal of content first before writing a successful prompt.
- Understand the difference between prompt writing and command stacking.
- Use AI for idea generation and regeneration.
- Develop UX writing scenarios to learn from AI content using specific terms.

AI Characteristics That Are Important to UX Writers

In the previous chapter, we mentioned how several developments in artificial intelligence (AI) have led to automated content creation and the need for UX writers to check and further develop "hybrid content." These developments in AI are useful because they add to a UX writer's toolbox by the following means:

DOI: 10.4324/9781003274414-15

- **Classification:** AI can classify different units in the text. For example, as you learned in the last chapter, AI can find out the intent of users, identify entities (such as "Mexican food"), and use these to create new output. Similarly, it can classify words and identify keywords in texts.
- **Dynamic learning:** AI can learn from new and refined input from a user. Similarly, AI can learn from a UX writer who modifies a request and asks AI to take in specific data and augment its output based on the new specifications of the UX writer.
- **Input of data:** Large amounts of data are analyzed by AI and inputted in training sets so they can be used to create new content, such as text, images, and music.
- **Idea generation:** Natural language generation uses contextual understanding of previous interactions from prompts to formulate a human-sounding conversational response.

As you can see, there are a lot of components to generative AI that are helpful to UX writers. Large language models of AI can do many things at once:

- Automate repetitive tasks.
- Generate and augment content quickly in different styles and genres of writing.
- Accelerate writing processes by bringing in new perspectives.
- Help brainstorm ideas when a writer is stuck.
- Analyze user data.
- Translate complex ideas and language into simpler language for different levels of understanding by an audience.
- Improve content by shortening or translating copy.
- Bring in new information and ideas by taking input from various sources (websites, data you provide) and incorporating this into its output. (Amer, 2023)

Through understanding the above capacities of AI, UX writers can develop scenarios to use AI to develop content that is more specific to their audience.

Furthermore, AI plugins in composing environments such as word processors, web search engines, and even design programs will integrate and augment the work processes of UX writers. For example, rather than edit a title manually, a UX writer may make use of an AI bot in their composing interface and simply write a text prompt and ask the AI to "make the title more conversational," and the AI will rewrite the title and insert a newer, more conversational manner. Alternatively, a design program may feature a text-to-image option, and a UX writer may ask, "create a prototype interface for a fitness app," and the AI will create an interface prototype with copy text, images, and an app interface. As you can tell, such plugins can speed up the work of UX writers by letting them quickly generate ideas with AI as a companion in their composing environment. For this reason, it is also important to know how to create good prompts.

How to Cook Up Good AI Prompts: Using the 6W and 1H Method to Frame AI Prompts

According to Rollins (2023), developing AI-written content is similar to creating cooking recipes. In writing with AI, you will need to know how to create good recipes that consist of ingredients that AI will understand and translate to your specific content and audience. Similar to cooking recipes, you will want to know what ingredients are needed to create a tasty dish for your specific audience (in our analogy, the "dish" stands for UX content, of course). Similar to cooking, any UX writing task requires you to first ask yourself which ingredients are needed to help create good content that resonates with your audience. And, similar to cooking, where you may have a specific image and taste in mind for the dish, you have to start with a broad idea of the desired outcome (or output) for the UX content you want AI to generate. Hence, using the output-first method of envisioning desired output and then getting more specific with the 6W and 1H questions can help you consider the nuts and bolts of writing a strong prompt for AI:

- **What:**
  - What is the desired *output* that I want from AI?
  - What *action* do I want the AI to perform?
  - What *topic/concept* am I asking AI to write about?
  - What *genre* is my UX content?
    - What is the *structure* of that genre?
    - What *components* need to be included?

- **Who:** Who is my *audience/user*?
- **Which:** Which UX *techniques* do I want to include?
- **Where:** *Where* will the user use this content? *Where* can I find additional resources/URLs that I can tell AI to take as input?
- **Why:** *Why* is it important for the user to know about this topic/concept?
- **When:** *When* will the user be using the content?
- **How:** *How* will the user make use of this content? *How long* is the content supposed to be?

Based on your answers to these 6W and 1H questions, you will need to think about how you will write your prompt so AI can generate the specific content that you need. Again, it is important for UX writers to understand that they need to humanize content written by AI by checking the accuracy of the work done by AI and adding their own special human touch.

Going through the above questions will let you specify your prompt for maximum effect. For example, you will want to specify which output you want from AI. If you are looking to create a blog post, cover letter, or email, you will want to specify that. Next, you will want to specify the AI to perform a particular action. For example, you can ask it to say "write" about something or "suggest" if you are looking for ideas. If you are generating, say, AI output for written text versus generating an interface or an image, your prompt will be different. Or if you are writing about a specific topic, such as "intercultural dimensions in UX" or "first-person-shooter video games," you will want to define your specific topic so the AI knows what direction you want it to go. Likewise, if you are asking AI to generate content in a specific genre, you will want to specify that (e.g., if you want AI to generate a social media post versus giving you SEO keywords).

Furthermore, you will also want to specify the structure and components of your content. For example, you may specify AI to write a social media post on a specific topic and ask AI to include "an opening hook, 2–3 lines of background information, and a call to action" for users to contact you. Specifying your audience is also good here (e.g., content for a technical, professional audience is different than that for teens). If you want AI to use a specific UX technique, you can specify that too (e.g., "write a journey map" or "create a persona based on characteristics of x, y, and z"). Other dimensions may also be good to think about when you are starting to write your prompt (e.g., you can feed textual data to AI and ask it to use this data as input for generating its output, or you can specify where/why/when, or how users will need to use the content). In other words, answering the 6W and 1H questions can thus help you write a better prompt for AI to target and write specific content. Again, AI can do the heavy lifting, but UX writers will still need to add and check content for accuracy and humanize the content output by rewriting it with their unique insights and perspectives.

Specific Prompt Commands for AI

Now that you know some good guiding questions to help you think about writing a better AI prompt recipe, you will want to know the specific `commands` for writing AI prompts. Think of AI commands as consisting of you asking AI to perform a specific action and deliver a desired output. The rest of the commands can modify that output and thus get you closer to what you need.

- **Action:** First, tell AI what *action* you want it to perform (e.g. `"write,"` `"summarize,"` `"suggest,"` `"incorporate,"` `"rewrite,"` `"rephrase,"` `"shorten,",` and `"analyze."`
- **Output:** Second, tell AI which *output* you want it to generate for you. Think of output as a *genre/structure* (e.g., "cover letter", "social media post", "recruitment email", etc.) that features *components* you specifically define.

  - For example, you can ask AI: `please write a social media post with an opening hook, 2-3 lines of background information, and call to action`. AI will then write your social media post with that exact structure of an opening hook, background info, and an end with a call to action.

Below are modifiers (that the AI will take as input) that help modify its output. These are, in other words, quite handy to narrow down what you are looking for in particular.

- **Topic:** Which topic do you want the AI to write about? For example, `please write a social media post about pet adoption with an opening hook, 2-3 lines of background information, and a call to action.`
- **User:** Define the user/audience for the content. For example, `please write a social media post about pet adoption with an opening hook, 2-3 lines of background information, and a call to action for an older audience.`
- **Dataset:** Which data do you want the AI to take into consideration as input (a specific app design, or you could provide specific source information like a website/give URL)? For example, `please write a social media post about pet adoption with an opening hook, 2-3 lines of background information, and a call to action for an older audience using text sample from http://www.aarp.org).` AARP.org (formerly known as the American Association of Retired Persons) is an organization that caters to Americans who are over 50 years old.
- **Tone of voice/style:** Define which type of tone of voice/style you want AI to write (`"folksy"`, `"friendly"`, `"conversational"`, `"witty"`, `"serious"`, etc.). For example, `"please write a social media post about pet adoption using a conversational tone,"` and so forth).
- **Technique:** Specify UX techniques from design thinking or UX that you want the AI to use to write content (such as `"persona,"` `"journey map,"` and `"AIDA - Attention, Interest, Desire, Action Framework"`). For example, `"please create a journey map for a pet adoption app for older audiences"`.

- **Format:** Asking AI to format the content in a specific way (e.g., output in tabular format, table, or just plain text). For example, "`please create a journey map for a pet adoption app for older audiences using a table format`". AI will create a table with your information.
- **Parameters/Tokens:** Length or requirement of content is often expressed in the number of words or tokens (1–2 tokens per word). For example, "`please create a journey map for a pet adoption app for older audiences using 500 words only`".
- **Temperature:** You can ask AI to be more strict in its logic or to be more creative by setting the temperature from 0 = strictly facts to 1 = creative with facts. Modifying the AI's temperature can not only lead to more creativity but also less accuracy. Hence, by telling the AI to "`set temperature to 1,`" you are asking the AI to get more creative and be less strict in its application of strict logic models.

As you can see, these commands are very important as they allow you to tell the AI what you specifically need, ranging from what actions you want the AI to take to what output you want and which input to use in creating that output.

You may have a pretty good idea of what you can do in your AI prompts using the 6W and 1H questions and the commands for AI prompts. An important aspect of working with AI is conceptualizing how it works at different stages of UX writing. First, writing AI prompts is recognizing when you have a good recipe and reusing that recipe. The command stacking strategy can help you quickly develop your repertoire in situations that are similar but where you want to make changes.

- **Prompt reuse:** You can create a prompt that works well and reuse it with different input variables. As mentioned earlier, seeing text prompts as recipes helps you reuse them to create new, effective content in a quick manner.

- **Command stacking:** You can also stack new commands by asking AI in follow-up commands to modify them in specific ways. You are then using command stacking since you are using the recipe but adding more commands and asking the AI to be more precise in its output. For example, you can ask AI to "`create a journey map for a fitness app`" and then ask a follow-up request "`now design it specifically for a teenager`", and it will write the journey map for the persona of a teenager.

What is interesting for UX writers is the ability of AI to incorporate UX techniques. Because AI can learn from the new input you give it, UX writers can also ask AI to incorporate techniques from UX design. For example, in your prompt, you can write "`generate a journey map,`" and it will generate a journey map for you. Alternatively, you can ask AI to "`generate a script for a fitness app, using the design thinking stages of empathize, ideate, prototype, and test,`" and ask the bot to write about each stage to generate ideas for, say, a fitness app. Seeing as these specific UX techniques are commonly used, below we will describe some common scenarios for UX writers when using AI.

Idea Generation, Regeneration, Suggestions, and UX Techniques

As it's been said many times, AI can be very helpful for UX writers (Thalion, 2023). AI can help augment the work of UX writers by helping to generate ideas, give suggestions, regenerate work for a fresh perspective, or even employ specific UX techniques in its output:

- **Idea generation:** AI is a great thinking tool for UX writers to just get some new ideas when they are stuck. Some jokingly call this "prompt paralysis"—i.e., when you are stuck and want to get the ball rolling (Rollins, 2023). Since AI uses a large learning model trained on many gigabytes of data, it can also give you suggestions when you just want to get started with some ideas. For example, you can ask it to write a poem, a cooking recipe, or even a story (`"once upon a time,...."`).
- **Regeneration:** AI can also regenerate content, so it is to the UX writer's liking! This can help create a different text output for the same prompt. Using regeneration also helps if, say, you ask the AI to `"write a call to action"` (a sample output may be "Sign up now to this online newsletter") and then ask it to `"now make it shorter"` (it may then output "Sign up").
- **Suggestions:** AI can also offer suggestions in case you want to get a lay-of-the-land feel. For example, it can suggest different usability testing techniques for a particular situation. A UX writer could then pick this information to make a choice of which usability technique to use.
- **UX techniques:** AI can also make UX-specific recommendations by incorporating personas, journey maps, wireframes, and content calendars into its output. You can pretty much ask for any of these specific UX, design thinking, or content strategy techniques.

As a UX writer, your work will go through the stages of design thinking. Using AI prompts during this process can therefore be very advantageous. Below, we give some scenarios for incorporating AI in your UX writing process using the well-known design thinking stages of empathize, ideate, prototype, and test. We note the importance of UX writers using the HEAT model introduced in the previous chapter to always check the work of AI and humanize hybrid content.

Empathize Using AI: Conducting Preliminary User Research

Generative AI can help you do user research. It can do this by not only suggesting helpful strategies for conducting user interviews but also by analyzing user data online and taking in the data. This can be really helpful if you want to get some preliminary inspiration when you are starting out with user research. It can also give you tentative ideas on how to conduct a user interview and what typical users you can expect.

Tips for user interviews and typical user demographic data:

- **User interviews:** Suggest 10 user interview questions for [insert your design idea, such as an app] to test usability and overall satisfaction with UX.
- **Ask demographic data:** What is the typical demographic for [insert your design idea, such as an app].

Analysis of textual data can be very useful for UX writers. For example, you can provide AI with regular text or existing consumer review data and ask it to analyze the user reviews on a particular product. This can be great to identify pain points or gauge the reactions of users informally using sentiment analysis or keyword classification.

Tips for user review data analysis:

- **Analyze textual data:** Analyze the text below for the 10 most common keywords [Insert your text or URL that you want analyzed].
- **Identify pain points:** Analyze [website address] and identify the top ten pain points in user reviews.

- **Sentiment analysis:** Analyze [user reviews] using sentiment analysis and classify whether people are overall more negative or positive about the product.

  - In a follow-up prompt, ask: Please identify the top ten negative and positive sentiments of users.

- **Keyword analysis:** What are the top ten keywords that reoccur in user reviews?

Using the above prompts, you have now identified common pain points, whether or not people speak positively or negatively about your product/service/design using sentiment analysis, and identified and classified the top ten negative and positive user experiences as well as recurring keywords (which give you a sense of how people discuss, in general, about the product). This is quite handy if you want to present improvements for UX on an existing product. Again, in keeping with the HEAT model, you will want to cross-check those user reviews, interview actual users, and consult with others in your company to check for the accuracy of the answers provided by AI.

Next, to doing this type of user research, you can also empathize with users by creating a persona and using that to analyze a specific website. This is especially helpful if you want to get insights and perspectives different from your own and cultivate a listening mindset, which is important for UX writers.

Tips for generating user personas:

- **Generate a persona:** Generate a user persona for a busy mom who lives in the city, doesn't like to cook, but wants her kids to eat healthy.

  - In a follow-up prompt, ask:
    Write this in the format of a user persona
  - You can then use this persona to analyze a website or product:
    Now use this persona to analyze [provide the website or text you want to be analyzed through this persona's lens].

AI will write a full user persona and provide a name, age, occupation, location, demographic data on what the mom's characteristics are (busy, attentive, etc.), goals, challenges, behaviors, attitudes, and pain points. It can also generate potential solutions for the persona (e.g., to eat healthy using meal delivery options, pre-made meals, etc.). And the great thing is, you can use this persona and ask the AI to analyze just about any website or service online and give you feedback from that persona's perspective. This can be quite handy to identify any blind spots from the perspective of that persona (since they are not the designer but the user). As we said in previous chapters on personas, however, it's critical that you verify that the data used to create the persona are based on real research you've compiled and that the persona isn't "fiction."

There are many other ways to empathize with users and stages in design thinking, such as creating customer journey maps, empathy maps, stakeholder maps, or using Activities, Environment, Interaction, Objects, and User (AEIOU). Plugging any of these terms into a prompt will for sure also help you define different components of the user experience and identify pain points and different channels and structures a user may interact with.

Tips for incorporating other design thinking techniques:

- **Generate customer journey map:** Develop an app similar to Miro and write a customer journey map for typical users.

- **Generate empathy map:** Generate an empathy map for a fitness app using hear, think & feel, hear, see, say & do, and identify pains and gains of typical fitness users.
- **Analyze UX content using the AEIOU framework:** Analyze the fitness app using the AEIOU framework and output it in a tabular format with rows for activities, environment, interaction, objects, and users.

As you can see from the above prompts, it is easy to incorporate specific design thinking techniques by naming the output you want. Again, it is important to verify your results through actual user testing, as AI can give you an idea of what a typical audience may do, but a UX writer will need to actually validate the ideas and test them with actual human users to successfully empathize with them. In other words, AI can give you inspiration, but you will still need to test it out with actual humans.

Define Using AI: Competitors, Common Pain Points, and Design Brief

Besides empathizing with your user and getting a sense of this in collaboration with AI, you can also use AI to define the problem space for your UX content. This can be helpful if you have to figure out what specific gap or issue you want to solve for your users and want to write a design brief.

Tips for defining and identifying problems and writing a design brief:

- **Competitor analysis:** Do a competitor analysis of the top five fitness apps.
- **Identify common pain points:** Identify common pain points in the top five fitness apps.
- **How might we question:** Using the "how might we" framework from UX, how might we improve common pain points for these fitness apps and create a new app.
- **Design brief:** Now write a design brief for this new app with the sections client/company, project goals & objectives, target audience, budget, schedule, and key deliverables.

Using the define stage, you have now done a competitor analysis, identified a gap in the market, and come up with a solution using the how might we questions. You have also started thinking ideas for a design brief, taking into account the preceding variables. There are, of course, many other techniques in design thinking that you can bring to this stage, but this will give you an idea of the types of things you can do. Also, note that you can use the sentiment analysis and keyword analysis from the empathize stage in your competitor analysis to dive in more deeply as to why or how specific pains could possibly exist. Again, a UX writer will need to bring the HEAT model into this discussion by checking to see if there is an actual need and cross-checking the work of AI.

Ideate Using AI: Page Layout, User Interface, and Wireframe

Next to defining your problem space, AI can also help generate ideas for an actual design. You can ask it to do many things, such as creating page layouts, generating ideas for a user interface (UI) or user experience, and even asking AI to suggest preliminary wireframing.

- **Get ideas for page layout:** Suggest a page layout for a fitness app.
- **Get ideas for UI/UX:** Suggest UI and UX for the fitness app.
- **Get ideas for a wireframe:** Suggest a wireframe for the fitness app using a tabular format.

For example, AI may recommend a wireframe and give you a table with text ideas on the navigation bar, a search bar, a main page design, informational nodes, and content in each node. Such ideas can be really good to learn from when you are in the stages of thinking through your design.

Prototype Using AI: User Flows, Design Systems, Copy

In addition to getting ideas for UI wireframes, you can also ask AI to prototype more fleshed-out designs. This can be quite handy if you are in the stages of prototyping and mocking up your content. As you saw in the preceding section, AI can suggest UI/UX, but it can also generate user flows, describe design systems, generate and optimize copy for websites or apps, and translate it to different audience levels.

- **Generate user flows:** Suggest a user flow for a typical fitness app from onboarding to becoming a regular user of said app.
- **Describe design systems:** Describe the top five design systems used for the accessibility of content.
- **Generate copy for an existing website/app:** Generate copy for the fitness app using an opening hook, 1–2 lines on what we do, and a call to action to sign up.
- **Optimize already-created copy (such as a title):** Now optimize the title of my content.
- **Shortening content:** Shorten the title, please.
- **Explaining difficult-to-understand terminology:** Now simplify the text so a five-year old can understand it.
- **Translate text into a different language:** Please translate the text from English to German.

The above commands are not an exhaustive list of what you can do to prototype content, but they will give you a sense of what you can do when you are prototyping content. The ability to optimize and translate language can be quite handy, specifically if you are wondering if the words you have written translate to an audience with a different level of understanding or language. For example, you may have call-to-action copy on your website such as "Please sign up now for our fitness app!". You can ask AI to optimize it, and it will most likely give a shorter "Please sign up now!". You can then ask to shorten it, and it will give you "sign up," and so forth.

This is also ideal if you want to quickly find out if a word will fit on a button by asking AI to translate or simplify or shorten the text. If you are using a text-to-image program (such as Midjourney or DALL-E), your output may be in visual format as you create visual components. Please see the "Create Images and User Interfaces Using AI Prompts" section below for specific prompts on this.

Test Using AI: Usability Tests, Interviews

You can get recommendations for AI when testing your content. These can provide inspiration on how to conduct usability tests, write scripted questions, and give you insights you may not have had before.

Tips for usability testing/user interviews:

- **Find a proper usability testing method:** Please recommend 10 usability testing methods for the fitness app.

- **Writing usability questions:** Please recommend 10 usability questions for a fitness app that focus on UX.
- **Asking about user experience:** Please recommend 10 interview questions that focus on overall UX.

The use of AI to develop tests for UX content can be quite useful, especially in combination with more specific details as to what you are trying to accomplish. Plugging in commands for specific components that you want to focus on, such as the emotional states of the user (anticipation, flow, pains, and gains), typical user behaviors, and so forth, can be quite helpful to give you insight as to what your user may do or not do. Again, it is important to validate by conducting actual user testing, as AI can only give you ideas for that process but not the flesh-and-blood experience of users.

Other General Tips

Apart from performing all the design thinking activities, you can also ask AI to identify developments in the field of UX, identify trends, or help suggest how to optimize your workflow.

General tips to keep up to date with UX and best practices:

- **Find UX tools to learn:** Tell me about the top ten UX tools that are helpful with prototyping.
- **Identify trends in UX:** Please identify the top ten trends in UX that are important now.
- **Ideas to optimize your workflow:** Please identify the best way to hand off design files to engineers.

Create Images and User Interfaces Using AI Prompts

Next to generative AI generating ideas in textual format, you can also ask text-to-image AI such as Midjourney or DALL-E to generate images for you using text prompts. This can be quite interesting, especially if you want to ask it to design a typical UI for a project (like an app) you have in mind. A lot of this can get overly technical, so below is just a simplified version of what you can do with a text-to-image generator using some simple commands. Text-to-image AI like Midjourney uses the **/imagine** command to start a prompt. The below examples are therefore more specific to Midjourney but can also apply to other text-to-image AI programs (Nielsen, 2022). For UX writers, it is probably a great idea to pass inspiration to AI in the form of an image that is similar to a UI or project they are doing.

Tips for passing an image/design for image inspiration to AI:

- **Using an image/URL for inspiration:** /imagine a fitness app using http://www.yourwebsite.com/user-interface-image.jpg as inspiration

This may be a very handy use of a prompt since you can take inspiration from a particular design and ask AI to create a similar UI. This works especially well if you also know how to modify it with the below commands.

Tips for specific image weight:

- **Giving more weight to image:** /imagine a fitness app using http://www.yourwebsite.com/user-interface-image.jpg --iw:4
  - The higher **the number,** the more weight is given to the AI to be exactly like the provided image.
- **Giving more weight to text elements:** /imagine a photo with a mother::3 baby::2 and a dog::1
  - The higher **the number** after the **double colon** (::), the more prominent the element will be in the image. Here, the mother will be the largest, followed by the baby, and then the dog.

As you can tell, using weights for elements can help you tell the AI what to prioritize in visualizing components. And you can get as specific as you like by, for instance, specifying a UI with white text on blue buttons.

Tips for styling your design:

- **Stylize:** /imagine UI app using an image of an athlete that is photorealistic and uses the colors blue, white, and red for the interface.
  - **Genre:** /imagine UI app for a fitness app using cyberpunk style.
  - **Artist name:** /imagine UI app in the style of Andy Warhol.
  - **Stylize:** /imagine UI fitness app --s 600.
    - The higher the amount after the - -s command, the more stylized the design becomes.

Tips for setting the resolution, width and height, realism, and filtering out elements:

- **Resolution:** /imagine UI fitness app 8k.
- **Set width and height size in pixels:** /imagine user interface fitness app --w 600--h 300.
- **Realism:** /imagine user UI app ultra photoreal.
- **Filtering out image components:** /imagine http://www.website.com/peasandcarrots.jpg --no peas.
  - The −no command ensures that you filter out a visual component that you do not want in the image, which in this case is peas.

Of course, the commands we suggested here are definitely not an exhaustive list of all commands for AI. However, this chapter can give you ideas on how to incorporate AI into UX processes. By using AI and command prompts, you can quickly generate content and get new ideas. By using AI plugins, UX writers can also incorporate AI into their work processes to help suggest or recommend new ideas in various compositing environments. Lastly, due to the fast developments in generative AI, it is important to always check the work it does and to humanize its content. AI is not perfect and always needs to be checked by a UX writer.

Conclusion

As you can tell, there is a lot that can be done if you know how to engineer prompts for generative AI. Similar to a cooking recipe, you can reuse prompts for different projects and modify them to the specific tastes of your audience. Using the 6W and 1H questions and asking yourself what the *output* is first, you can then start to think about what you are trying to accomplish. Asking AI to perform a specific *action* (such as generate, suggest, or shorten) and specifying your *output* (in the form of a specific UX genre, such as a blog post, social media post, etc.) are the main steps. You can then use different *modifiers* to be more specific and define the style it is written in, what components should be a part of it, and so forth. In this chapter, you have also learned how you can use AI prompts for some common UX goals and specific design thinking stages, from empathizing to testing. From this, you have learned to ask AI to create ideas for journey maps, personas, and some other UX techniques. You have also learned about passing data and images to AI as prompts to create new UX content. As always, AI can only give you insights and new ideas, but you need to employ the HEAT method we discussed in Chapter 11 to humanize AI and tailor what you are making for the specific audience you have in mind. While AI may give ideas, it does not have a clear idea of what your audience wants, and as a UX writer, you will still need to test your ideas with an actual human audience. Likewise, you will want to always check the output of AI, seeing as it may generate inaccurate content. For this reason, it is important that UX writers see their own role in the process of using AI to create accurate and audience-friendly content.

Chapter Checklist

- *Use the 6W + 1H questions to specify the desired content you want from generative AI.*

    - What output? Action? Topic? Genre?
    - Who is the audience?
    - Why is the content important?
    - Which UX techniques to employ?
    - Where will the content be used?
    - When will the content be used?
    - How will the content be used?

- *Create specific commands for the AI to produce stronger content.*

    - Tell the AI what you want it to do (action).
    - Tell the AI what you want it to generate (output).

- *Apply prompt reuse and command stacking to enhance content.*

    - Create prompts that work well and reuse them with different input variables.
    - Ask AI with follow-up commands to modify content.

- *Perform design thinking alongside AI.*

    - Empathize using AI: Conduct user research, analyze user data, and generate personas.
    - Define using AI: Identify problems, market competitors, and create a design brief.
    - Ideate using AI: Generate ideas for page, UI, and wireframe design.
    - Prototype using AI: Create user flows, design systems, and copy.
    - Testing using AI: Design usability tests and conduct user interviews.

- *Create images and UIs using AI prompts.*

    - Specify image requirements such as weight, style, resolution, size, etc.

Discussion Questions

1 What ethical considerations do you make when creating AI prompts?
2 How do you plan to credit generative AI in your design projects?
3 Who should own the copyright to your prompts and AI responses?
4 Where can you save reusable or stackable prompts?

Learning Activity

Now is the time to play! Use a generative AI like ChatGPT and apply the recommended prompts from the chapter to create a set of UX content for a mobile app of your choice. Note the power and limitations of these prompts. Discuss with a friend about the issues you ran into when generating the content and how you worked around them.

References

Amer, M. (2023). Generative AI with Cohere. https://txt.cohere.ai/generative-ai-part-2/

Nielsen, L. (2022). An advanced guide to writing prompts for Midjourney (text-to-image). https://medium.com/mlearning-ai/an-advanced-guide-to-writing-prompts-for-midjourney-text-to-image-aa12a1e33b6

Rollins, Darby. (2023). *Create content with AI*. The AIAuthor.

Thalion. (2023). How to use ChatGPT for UI/UX design: 25 examples. https://blog.prototypr.io/how-to-use-chatgpt-for-ui-ux-design-25-examples-f7772bea3e70

Glossary of Key Terms

Agile   An iterative, continuous, and cyclical approach to collaborative project management. See Chapter 3 on practicing agile collaboration.

Analytics   Curated, processed, and visualized datasets to understand audiences and decision-making. See Chapter 8 on using data analytics to understand users.

Artificial intelligence   The theory and development of computing systems that augment human intelligence, including visual perception, speech recognition, decision-making, and language translation. See Chapter 2 on the influence of artificial intelligence in content design and Chapters 11 and 12 for specific recommendations on using generative AI.

Auditing   Activities involving the analysis and evaluation of the content environment (site, partner content, parent and sister sites, etc.). See Chapter 2 on the content lifecycle.

Bounce rate   A metric that expresses how quickly a user leaves your website. See Chapter 8 on adopting performance metrics.

Content lifecycle   An iterative process that shapes the workflow of content creation, distribution, performance measurement, and improvement. The content lifecycle includes the following stages: auditing and analyzing, strategizing, planning, creating, and maintaining. See Chapter 2 on the UX writing process.

Content management systems   Tools or platforms that support collaborative composing, single-sourcing workflow, and asset management. See Chapter 2 on content management in the content lifecycle.

Contextual inquiry   Activities involving the watching, listening, and talking with users as they do their work in their own work settings. See Chapter 4 on contextual inquiry activities.

Definition activities   involving the unpacking and translating of user research findings into compelling, actionable insights. See Chapter 5 on definition activities.

Design thinking   A mindset and methodology that leverages individual experiences and collective expectations to create user-centered solutions. Design thinking specifies five distinctive phases: empathy, definition, ideation, prototyping, and testing. See Chapter 2 on the history of design thinking.

Empathy   Understanding users' behaviors and needs in the contexts of their work and life. See Chapter 4 on empathy activities.

Fidelity   The distance between the look and feel of the prototype and the final design. See Chapter 6 on creating low-fidelity and high-fidelity prototypes.

Human-centered design   A principle for creating content based on empathy with users, which requires designers to step outside of their regular worldview, background, and personal values. Consistent with ISO 9241-210:2019–Ergonomics of human-system interaction — Part 210: Human-centered design for interactive systems. See Chapter 2 on the influence of human-centered design in UX writing.

Ideation   Generating radical solutions and alternatives based on defined problems and opportunities. See Chapter 6 on ideation activities.

Interviews   A method for goal setting, collecting users' priorities and needs, and understanding business goals. See Chapter 4 on interview activities.

Iterative process   A cyclic approach to design through constant prototyping, testing, and refining a solution. See Chapter 2 on the iterative nature of UX writing.

Journey maps   A visual used to envision and track user interactions with a system. See Chapter 5 on journey mapping activities.

Key-performance   indicator A goal specified by an organization that can be measured and expressed through a metric. See Chapter 8 on applying key-performance indicators to measure content performance.

Mental models   A way to explain how humans learn to interact with interfaces and figure out unfamiliar situations. See Chapter 2 on understanding users' mental models.

Microcontent   Short-form content that helps users to quickly digest traditionally long-form content. See Chapter 9 for microcontent examples.

Microcopy   Short texts on websites, digital applications, and interactive interfaces that provide information or instructions to users. See Chapter 9 for microcopy examples.

Minimum viable product   The smallest product release that successfully achieves its desired outcomes. See Chapter 4 on the purpose of building the minimum viable product.

Multichannel   Multiple channels (e.g., web, app, phone, and chatbots) operating independently on a user journey. See Chapter 7 on the multichannel approach.

Net promoter score   A score between -100 and 100, generated from having the percentage of people that like a product or service and would promote it and subtracting the percentage of people that would actively discourage people from using. See Chapter 8 on using net promoter score to measure how users feel about a product.

Omnichannel   Multiple channels operating in an integrated and seamless manner to enhance users' journey. See Chapter 7 on the omnichannel approach.

Persona   A composite based on research that provides an archetypal image of different types of users or audiences for a product.

Portfolio   A collection of selected artifacts that showcase the strengths and potential of a UX writer. See Chapter 10 on creating a UX writing portfolio.

Prototyping   Getting ideas into the real world through material or digital means. See Chapter 6 on prototyping activities.

Single-source authoring   A structured writing process that applies a modular content creation and management approach to design. See Chapter 2 for single-sourcing examples.

Surveys   A way to gather large amounts of information by asking a predefined group of people to complete a research questionnaire. See Chapter 5 for sample applications of surveys.

System usability scale (SUS)   A standardized questionnaire for gathering users' reactions to the usability and learnability of a system. See Chapter 7 on the uses of the SUS.

Testing   Getting user feedback on the designed solution. See Chapter 7 on testing activities.

Think-aloud protocol (aka usability testing)   A method to build empathy with users by asking a representative user to perform a normal task with a product and to say out loud what they are thinking as they perform that task. See Chapter 4 for examples of thinking-aloud protocols.

Touchpoint options   in which a user is given to receive information about their interactions with a system. See Chapter 7 for touchpoint examples.

Tracking   Activities involving the trailing of content performance and measuring success. See Chapter 8 on tracking activities and technologies.

Usability   A measure to describe how well a user interacts with a design to achieve desired goals effectively, efficiently, and satisfactorily. See Chapter 4 on usability testing.

User experience (UX)   All aspects of the human experience when interacting with an interface, system, service, and organization. See Chapter 1 for an introduction to user experience.

UX writing   UX writing, or writing for user experience, is concerned with the integrative experience between the user and product/service as it is mediated by different content. See Chapter 1 for a more detailed definition.

Index

Note: **Bold** page numbers refer to tables and *italic* page numbers refer to figures.